Introduction to Digital Music with Python Programming

Introduction to Digital Music with Python Programming provides a foundation in music and code for the beginner. It shows how coding empowers new forms of creative expression while simplifying and automating many of the tedious aspects of production and composition.

With the help of online, interactive examples, this book covers the fundamentals of rhythm, chord structure, and melodic composition alongside the basics of digital production. Each new concept is anchored in a real-world musical example that will have you making beats in a matter of minutes.

Music is also a great way to learn core programming concepts such as loops, variables, lists, and functions, *Introduction to Digital Music with Python Programming* is designed for beginners of all backgrounds, including high school students, undergraduates, and aspiring professionals, and requires no previous experience with music or code.

Michael S. Horn is Associate Professor of Computer Science and Learning Sciences at Northwestern University in Evanston, Illinois, where he directs the Tangible Interaction Design and Learning (TIDAL) Lab.

Melanie West is a PhD student in Learning Sciences at Northwestern University and co-founder of Tiz Media Foundation, a nonprofit dedicated to empowering underrepresented youth through science, technology, engineering, and mathematics (STEM) programs.

Cameron Roberts is a software developer and musician living in Chicago. He holds degrees from Northwestern University in Music Performance and Computer Science.

Introduction to Digital Music with Python Programming

Learning Music with Code

Michael S. Horn, Melanie West, and Cameron Roberts

Routledge
Taylor & Francis Group

LONDON AND NEW YORK

First published 2022
by Routledge
4 Park Square, Milton Park, Abingdon, Oxon OX14 4RN

and by Routledge
605 Third Avenue, New York, NY 10158

Routledge is an imprint of the Taylor & Francis Group, an informa business

British Library Cataloguing-in-Publication Data
A catalogue record for this book is available from the British Library

Library of Congress Cataloging-in-Publication Data
A catalog record has been requested for this book

ISBN: 978-0-367-47083-8 (hbk)
ISBN: 978-0-367-47082-1 (pbk)
ISBN: 978-1-003-03324-0 (ebk)

DOI: 10.4324/9781003033240

Typeset in Bembo
by codeMantra

Melanie dedicates her contribution of the book to the memory of her friend, Bernie Worrell, who taught her how to listen.

Mike dedicates his contribution to his wife, Diana Reed, and his children, Madeleine and Lucas.

Contents

Figures

Photo and illustration credits

George Folz (georgefolz.com) created original illustrations for
  this book featured in Interludes 2, 4, and 8.
The photograph at the beginning of Chapter 3 is by
  (unsplash.com/@jonathanvez)
The photograph at the beginning of Chapter 4 is by
  (unsplash.com/@markus_gjengaar)
The photograph at the beginning of Chapter 5 is by
  (unsplash.com/@jasmund)
The photograph at the beginning of Chapter 6 is by
  (unsplash.com/@efrenbarahona3)
The photograph at the beginning of Chapter 7 is by
  (unsplash.com/@pablodelafuente)
The photograph at the beginning of Chapter 8 is by
  (unsplash.com/@makuph)
The photograph at the beginning of Chapter 10 is by
  (unsplash.com/@didierjoomun)
The photograph at the beginning of Interlude 9 is by
  (unsplash.com/@halacious)
The photograph at the beginning of Chapter 11 is by deepsonic
  (flickr.com/people/73143485@N02). Creative
  Commons License creativecommons.org/licenses/by/2.0.

Foreword

When I was a kid growing up in Texas, I "learned" how to play viola. I put *learned* in quotes because it was really just a process of rote memorization—hours and hours of playing the same songs over and over again. I learned how to read sheet music, but only to the extent that I knew the note names and could translate them into the grossest of physical movements. I never learned to read music as literature, to understand its deeper meaning, structure, or historical context. I never understood anything about music theory beyond being annoyed that I had to pay attention to accidentals in different keys. I never composed *anything*, not even informally scratching out a tune. I never developed habits of deep listening, of taking songs apart in my head and puzzling over how they were put together in the first place. I never played just for fun. And, despite the best intentions of my parents and teachers, I never fell in love with music.

Learning how to code was the complete opposite experience for me. I was largely self-taught. The courses I took in school were electives that I chose for myself. Teachers gave me important scaffolding at just the right times, but it never felt forced. I spent hours working on games or other projects (probably when I should have been practicing viola). I drew artwork, planned out algorithms, and even synthesized my own rudimentary sound effects. I had no idea what I was doing, but that was liberating. No one was around to point out my mistakes or to show me how to do things the "right" way (at least, not until college). I learned how to figure things out for myself, and the skills I picked up from those experiences are still relevant today. I fell in love with coding.[1]

But I know many people whose stories are flipped 180 degrees. For them, music was so personally, socially, and culturally motivating that they couldn't get enough. They'd practice for hours and hours, not just for fun but for something much deeper. For some it was an instrument like the guitar that got them started. For others it was an app like GarageBand that gave them a playful entry point into musical ideas. To the extent that they had coding experiences, those experiences ranged from uninspiring to off-putting. It's not that they necessarily hated coding, but it was something they saw as not being for them.

In the foreword of his book, *Mindstorms: Children, Computers, and Powerful Ideas,* Seymour Papert wrote that he "fell in love with gears" as a way of helping us imagine a future in which children (like me) would fall in love with computer programming, not for its own sake, but for the creative worlds and powerful ideas that programming could open up. Part of what he was saying was that love and learning go hand in hand, and that computers could be an entry point into many creative and artistic domains such as mathematics and music. Coding can revitalize subjects that have become painfully rote in schools.

The process of developing TunePad over the past several years has been a fascinating rediscovery of musical ideas for me. Code has given me a different kind of language for thinking about things like rhythm, chords, and harmony. I can experiment with composition unencumbered by my maladroit hands. Music has become something creative and alive in a way that it never was for me before. Music theory is no longer a thicket of confusing terminology and instead has become a fascinating world of mathematical beauty that structures the creative process.

Melanie, Cameron, and I hope that this book gives you a similarly joyful learning experience with music and code. We hope that you feel empowered to explore the algorithmic and mathematical beauty of music. We hope that you discover, as we have, that music and code reinforce one another in surprising and powerful ways that open new creative opportunities for you. We hope that, regardless of your starting point—as a coder, as a musician, as neither, as both—you will discover something new about yourself and what you can become.

Michael Horn
Chicago, Illinois (July 2021)

Note

1 I was also fortunate to have grown up in a time and place where these activities were seen as socially acceptable for a person of my background and identity.

Acknowledgments

We are grateful to the many people who have helped make this book possible. We especially want to thank Dr. Amartya Banerjee who has anchored the TunePad development team. The TunePad project grew out of a collaboration with the EarSketch team at Georgia Tech that was initiated by Dr. Brian Magerko and Dr. Jason Freeman. We thank Dr. Nichole Pinkard, Dr. Amy Pratt, and the Northwestern Office of Community Education Partnerships. We thank the TIDAL Lab team at Northwestern University including Mmachi Obiorah, Wade Berger, Izaiah Wallace, Brian Andrus, Jamie Gorrson, Matthew Brucker, Lexie Zhao, Ayse Hunt, Kallayah Henderson, Cortez Watson Jr., Sachin Srivastava, and many, many others. We thank our community partners including the Evanston Public Library, the NAACP of DuPage County, the James R. Jordan Foundation, the Meta Media program at the McGaw YMCA, the Hip-Hop FIRM, EvanSTEM, the Center for Creative Entrepreneurship, Studio 2112, the James R. Jordan Boys and Girls Club, Lake View High School and Marshaun Brooks, Lane Tech High School and Amy Wozniak, Gary Comer Youth Centers, and Chicago Youth Centers, Project Exploration, BBF Family Services, and the Museum of Science and Industry. Shout-outs to Marcus Prince and Sam Carroll who gave us insightful curriculum ideas, to Tom Knapp who contributed to TunePad's graphical design, and to the amazing interns we've worried with over the years.

Special thanks go to the people who gave input into the ideas and text of this manuscript including George Papajohn and Diana Reed. We also thank Joseph Mahanes, Abbie Reeves, and others who put up with us while we worked on this book.

TunePad was created by the Tangible Interaction Design and Learning (TIDAL) Lab at Northwestern University in collaboration with the EarSketch team at the Georgia Institute of Technology and with funding from the National Science Foundation (grants DRL-1612619, DRL-1451762, and DRL-1837661) and the Verizon Foundation. Any opinions, findings, and/or recommendations expressed in the material are those of the authors and do not necessarily reflect the views of the funders.

1 Why music and coding?

Welcome to *Introduction to Digital Music with Python: Learning Music with Code.* This book is designed for people who *love* music and are interested in the intersection of music and coding. Maybe you're an aspiring musician or music producer who wants to know more about coding and what it can do. Or maybe you already know a little about coding, and you want to expand your creative musical horizon. Or maybe you're a total beginner in both. Regardless of your starting point, this book is designed for you to learn about music and coding as mutually reinforcing skills. Code gives us an elegant language to think about musical ideas, and music gives us a context within which code makes sense and is immediately useful. Together they form a powerful new way to create music that will be interconnected with digital production tools of the future.

More and more code will be used to produce music, to compose music, and even to perform music for live audiences. Digital production tools such as Logic, Reason, Pro Tools, FL Studio, and Ableton Live are complex software applications created with *millions* of lines of code written by huge teams of software engineers. With all of these tools you can write code to create custom plugins and effects. Beyond production tools, live coding is an emerging form of musical performance art in which Information Age DJs write computer code to generate music in real time for live audiences.

In other ways, we're still on the cusp of a radical transformation in the way that we use code to create music. The history of innovation in music has always been entwined with innovation in technology. Whether we're

DOI: 10.4324/9781003033240-1

talking about Franz Liszt in the 19th century, who pioneered the persona of the modern music virtuoso based on technological breakthroughs of the piano,[1] or DJ Kool Herc in the 20th century, who pioneered hip-hop with two turntables and a crate full of funk records in the Bronx, technologies have created new opportunities for musical expression that have challenged the status quo and given birth to new genres. We don't have the Franz Liszt or DJ Kool Herc of coding yet, but it's only a matter of time before the coding virtuosos of tomorrow expand the boundaries of what's possible in musical composition, production, and performance.

1.1 What is Python?

In this book you'll learn how to create your own digital music using a computer programming language called **Python**. If you're not familiar with programming languages, Python is a general-purpose language first released in the 1990s that is now one of the most widely used languages in the world. Python is designed to be easy to read and write, which makes it a popular choice for beginners. It's also fully featured and powerful, making it a good choice for professionals working in fields as diverse as data science, web development, the arts, and video game development. Because Python has been around for decades, it runs on every major computer operating system. The examples in this book even use a version of Python that runs directly inside of your web browser without the need for any special software installation.

Unlike many other common beginner programming languages, Python is "text-based", which means that you type code into an editor instead of dragging code blocks on the computer screen. This makes Python a little harder to learn than other beginner languages, but it also greatly expands what you can do. By the time you get through this book you should feel comfortable writing short Python programs and have the conceptual tools you need to explore more on your own.

1.2 What this book is *not*

Before we get into a concrete example of what you can do with a little bit of code, just a quick note about what this book is *not*. This book is not a comprehensive guide to Python programming. There are many excellent books and tutorials designed for beginners, several of which are free.[2]

This book is also not a comprehensive guide to music theory or Western music notation. We'll get into the core ideas behind rhythm, harmony, melody, and composition, but there are, again, many other resources available for beginners who want to go deeper. What we're offering is a different approach that combines learning music with learning code in equal measure.

1.3 What this book *is*

What we will do is give you an intuitive understanding of the fundamental concepts behind both music and coding. Code and music are highly technical skills, full of arcane symbols and terminology, that seem almost designed to intimidate beginners. In this book we'll put core concepts to use immediately to start making music. You'll get to play with ideas at your own pace and get instant feedback as you bring ideas to life. We skip most of the technical jargon and minutiae for now—that can come later. Instead, we focus on developing your confidence and understanding. Importantly, the skills, tools, and ways of thinking that we introduce in this book will be broadly applicable in many other areas as well. You'll be working in Python code, but the core structures of variables, functions, loops, conditional logical, and classes are the same across many programming languages including JavaScript, Java, C, C++, and C#. After you learn one programming language, each additional language is that much easier to pick up.

1.4 TunePad and EarSketch

This book uses two free online platforms that combine music and Python coding. The first, called TunePad (https://tunepad.com), was developed by a team of researchers at Northwestern University in Chicago. TunePad lets you create short musical loops that you can layer together using a simple digital audio workstation (DAW) interface. The second platform, called EarSketch (https://earsketch.gatech. edu), was created by researchers at Georgia Tech in Atlanta. EarSketch uses Python code to arrange samples and loops into full-length compositions. Both platforms are browser-based apps, so all you need to get started is a computer (tablets or Chromebooks are fine), an internet connection, and a web browser like Chrome or Firefox. External speakers or headphones are also nice but not required. Both platforms have been around for years and have been used by many thousands of students from middle school all the way up to college and beyond. TunePad and EarSketch are designed primarily as learning platforms, but there are easy ways to export your work to professional production software if you want to go further.

1.5 A quick example

Here's a quick example of what coding in Python looks like. This program runs in TunePad to create a simple beat pattern, variants of which have been used in literally thousands of songs such as *Blinding Lights* by The Weeknd and *Roses* by SAINt JHN.

```
1  playNote(1)              # play a kick drum sound
2  playNote(2)              # play a snare drum sound
3  playNote(1)
4  playNote(2)
5  rewind(4)                # rewind 4 beats
6  for i in range(4):
7      rest(0.5)
8      playNote(4, beats = 0.5)  # play hat for a half beat
```

These eight lines of Python code tell TunePad to play a pattern of kick drums, snare drums, and high-hats. Most of the lines are **playNote** instructions, and, as you might have guessed, those instructions tell TunePad to play musical sounds indicated by the numbers inside of the parentheses. This example also includes something called a *loop* on line 6. Don't worry too much about the details yet, but the loop is an easy way to repeat a set of actions over and over again. In this case, the loop tells Python to repeat lines 7 and 8 four times in a row. The screenshot (Figure 1.1) shows what this looks like in TunePad. You can try out the example for yourself with this link: https://tunepad.com/examples/roses.

1.6 Five reasons to learn code

Now that you've seen a brief example of what you can do with a few lines of Python code, here are our top five reasons to get started with programming and music if you're still in doubt.

1.6.1 REASON 1: Like it or note, music is already defined by code

Looking across the modern musical landscape, it's clear that music is already defined by code. One of the biggest common factors of almost all modern music from any popular genre is that *everything* is edited, if not created entirely, with sophisticated computer software. It's hard to overstate how profoundly such software has shaped the sound of music in the 21st century. Relatively inexpensive DAW applications and the myriad ubiquitous plugins that work across platforms have had a disruptive and democratizing effect across the music industry. Think about effects plugins like autotune, reverb, or the ability to change the pitch of a sample without changing the tempo. These effects are all generated with sophisticated software. Production studios the size of small offices containing hundreds of thousands of dollars' worth of equipment now fit on the screen of a laptop computer available to any aspiring producer with passion, a WiFi

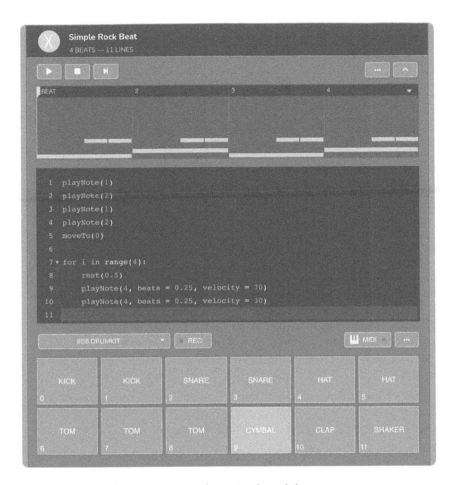

Figure 1.1 A TunePad program to play a simple rock beat.

connection, and a small budget. The reasons behind the shift to digital production tools are obvious. Computers have gotten to a point where they are cheap enough, fast enough, and capacious enough to do real-time audio editing. We can convert sound waves into editable digital information with microsecond precision and then hear the effects of our changes in real time. These DAWs didn't just appear out of nowhere. They were constructed by huge teams of software engineers writing code—millions of lines of it. As an example, TunePad was created with over 1.5 million lines of code written in over a dozen computer languages such as Python, HTML, JavaScript, CSS, and Dart. Regardless of how you feel about the digital nature of modern music, it's not going away. Learning to code will

Figure 1.2 Typical DAW software.

help you understand a little more about how all of this works under the hood. More to the point, it's increasingly common for producers to write their own code to manipulate sound. For example, in Logic, you can write JavaScript code to process incoming MIDI (Musical Instrument Digital Interface) data to do things like create custom arpeggiators. Learning to code can give you more control and help expand your creative potential (Figure 1.2).

1.6.2 REASON 2: Code is a powerful way to make music

We don't always think about it this way, but music is *algorithmic* in nature—it's full of mathematical relationships, logical structure, and recursive patterns. The beauty of the Baroque fugue is in part a reflection of the beauty of the mathematical and computational ideas behind the music. We call Bach a genius not just because his music is so compelling, but also because he was able to hold complex algorithms in his mind and then transcribe them to paper using the representation system that we call Western music notation. In other words, music notation is a language for recording the output of the composition process, but not a language for capturing the algorithmic nature of the composition process itself.

Code, on the other hand, is a language specifically designed to capture mathematical relationships, logical structure, and recursive patterns. For example, take the stuttered hi-hat patterns that are one of the defining characteristics of trap music. Here are a few lines of Python code that generate randomized hi-hat stutters that can bring an otherwise conventional beat to life with sparkling energy.

```
1  for _ in range(16):
2      if randint(6) > 1:           # roll the die for a random number
3          playNote(4, beats=0.5)   # play an eighth note
4      else:
5          playNote(4, beats=0.25)  # or play 16th notes
6          playNote(4, beats=0.25)
```

Or, as another example, here's a two-line Python program that plays a snare drum riser effect common in house, EDM, or pop music. You'll often hear this technique right before the beat drops. This code uses a decay function so that each successive note is a little shorter resulting in a gradual acceleration effect. Don't worry about how all of this works just yet. We'll walk you through the details as we go along.

```
1  for i in range(50):                          # play 50 snares
2      playNote(2, beats = pow(2, -0.09 * i))
```

What's cool about these effects is that they're *parameterized*. Because the code describes the algorithms to generate music, and not the music itself, it means we can create infinite variation by adjusting the numbers involved. For example, in the trap hi-hat code, we can easily play around with how frequently stuttered hats are inserted into the pattern by increasing or decreasing one number. You can think of code as something like a power drill; you can swap out different bits to make holes of different sizes. The drill bits are like parameters that change what the tool does in each specific instance. In the same way, algorithms are vastly more general-purpose tools that can accomplish myriad tasks by changing the input parameters.

Creating a snare drum riser with code is obviously a very different kind of thing than picking up two drumsticks and banging out a pattern on a real drum. And, to be clear, we're not advocating for code to replace learning how to perform with live musical instruments. But, code can be another tool in your musical repertoire for generating repetitive patterns, exploring mathematical ideas, or playing sequences that are too fast or intricate to play by hand.

1.6.3 REASON 3: Code lets you build your own musical toolkit

Becoming a professional in any field is about developing expertise with tools—acquiring equipment and knowing how to use it. Clearly, this is true in the music industry, but it's also true in software. Professional software engineers acquire specialized equipment and software packages. They develop expertise in a range of programming languages and technical frameworks. But, they also build their own specialized tools that they use across projects. In this book, we'll show you how to build up your own library of Python functions. You can think of functions as specialized tools that you create to perform different musical tasks. In addition to the examples we described above, you might write a function to generate a chord progression or play an arpeggio, and you can use functions again and again across many musical projects.

1.6.4 REASON 4: Code is useful for a thousand and one other things

As we mentioned earlier in this chapter, Python is one of the most powerful, multi-purpose languages in the world. It's used to create web servers and social media platforms as much as video games, animation, and music. It's used for research and data science, politics and journalism. Knowing a little Python gives you access to powerful machine learning and artificial intelligence (AI/ML) techniques that are poised to transform most aspects of human work, including in creative domains such as music. Python is both a scripting language and a software engineering platform—equal parts duct tape and table saw—and it's capable of everything from quick fixes to durable software applications. Learning a little Python won't make you a software engineer, just like learning a few guitar chords won't make you a performance musician. But it's a start down a path. An open door that was previously closed, and a new way of using your mind and a new way of thinking about music.

1.6.5 REASON 5: Coding makes us more human

When we think about learning to code, we tend to think about the economic payoff. You'll hear arguments that learning to code is a resume builder and a path to a high-paying job. It's not that this perspective is wrong, but it might be the wrong reason for *you* to learn how to code.

Just like people who are good at music *love* music, people who are good at coding tend to *love* coding. The craft of building software can be tedious and frustrating, but it can also be rewarding. It's a way to express oneself creatively and to engage in craftwork. People don't learn to knit, cook, or play an instrument for the lucrative career paths that these pursuits open up—although by all means those pursuits can lead to remarkable careers. People learn these things because they have a *passion* for them. Because they are personally fulfilling. These passions connect us to centuries of tradition; they connect us to communities of teachers, learners, and practitioners; and, in the end, they make us more *human*. So when things get a little frustrating—and things always get a little frustrating when you're learning any worthwhile skill—remember that just like poetry, literature, or music, code is an art as much as it is a science. And, just like woodworking, knitting, or cooking, code is a craft as much as it is an engineering discipline. Be patient and give yourself a chance to fall in love with coding.

1.7 The future of music and code

Before we get on with the book, we wanted to leave you with a brief thought about the future of technology, music, and code. For as long as there have been people on this planet there has been music. And, as long as there has been music, people have created technology to expand and enhance their creative potential. A drum is a kind of technology—a piece of animal hide stretched across a hollow log and tied in place. It's a polylithic accomplishment, an assembly of parts that requires skill and craft to make. One must know how to prepare animal hide, to make rope from plant fiber, and to craft and sharpen tools. More than that, one must know how to perform with the drum, to connect with an audience, to enchant them to move their bodies through an emotional and rhythmic connection to the beat. Technology brings together materials and tools with knowledge. People must have knowledge both to craft an artifact and to wield it. And, over time—over generations—that knowledge is refined as it gets passed down from teacher to student. It becomes stylized and diversified. Tools, artifacts, knowledge, and practice all become something greater. Something we call culture.

Again and again the world of music has been disrupted, democratized, and redefined by new technologies. Hip-hop was a rebellion against the musical status quo fueled by low-cost technologies like recordable cassette tapes, turntables, and 808 drum machines. Early innovators shattered norms of artistic expression, redefining music, poetry, visual art, and dance in the process. Inexpensive access to technology coupled with a need for new forms of authentic self-expression was a match to the dry tinder of racial and economic oppression.

It's hard to overstate how quickly the world is still changing as a result of technological advancements. Digital artifacts and infrastructures are so ubiquitous that they have reconfigured social, economic, legal, and political systems; revolutionized scientific research; upended the arts and culture; and even wormed their way into the most intimate aspects of our personal and romantic lives. We've already talked about the transformative impact that digital tools have had on the world of music in the 21st century, but the exhilarating (and scary) part is that we're on the precipice of another wave of transformation in which human creativity will be redefined by Artificial Intelligence AI and Machine Learning ML. Imagine AI accompanists that can improvise harmonies or melodies in real time with human musicians. Or deep learning algorithms that can listen to millions of songs and innovate music in the same genre. Or silicon poets that grasp human language well enough to compose intricate rap lyrics. Or machines with trillions of transistor synapses so complex that they begin to "dream"—inverted machine learning algorithms that ooze imagery unhinged enough to disturb the absinthe slumber of surrealist painters. Now, imagine that this is not speculative science fiction, but the reality of our world today. These things are here now and already challenging what we mean by human creativity. What are the implications of a society of digital creative cyborgs?

But here's the trick: we've always been cyborgs. Western music notation is as much a technology as Python code. Becoming literate in any suffi-ciently advanced representation system profoundly shapes how we think about and perceive the world around us. Classical music notation, theory, and practice shaped the mind of Beethoven as much as he shaped music with it—so much so that he was still able to compose many of his most famous works while almost totally deaf. Beethoven was a creative cyborg enhanced by the technology of Western music notation and theory. The difference is that now we've externalized many of the cognitive processes into machines that think alongside us. And, increasingly, these tools are available to everyone. How that changes what it means to be a creative human being is anyone's guess.

1.8 Book overview

We're excited to have you with us on this journey through music and code. Here's a short guide for where we go from here. Chapters 2 and 3 cover the foundations of rhythm, pitch, and harmony. These chapters are designed to move quickly and get you coding in Python early on. We'll cover Python variables, loops, which both connect directly to musical concepts. Chapters 4, 5, and 6 cover the foundations of chords, scales, and keys using Python lists, functions, and data structures. Chapters 7, 8, and 10 shift from music composition to music production covering topics such as the frequency domain, modular synthesis, and other production effects. In Chapter 9, we switch to the EarSketch platform to talk about how various musical elements are combined to compose full-length songs. Finally, Chapter 11 provides a short overview of the history of music and code along with a glimpse of what the future might hold. Between each chapter, we provide a series of short *interludes* that are like step-by-step tutorials to introduce new music and coding concepts.

A few notes about how to read this book. Any time we include Python code, it will be shown in a programming font like this:

```
playNote(60)
```

Sometimes we'll write code in a table with line numbers so that we can re-fer to specific lines. When we introduce new terms, we'll **bold** the word. If you get confused by any of the programming or music terminology, check out the appendices, which contain quick overviews of all of the important concepts. We'll often invite you to follow along with online examples. The best way to learn is by doing it yourself, so we strongly encourage you to try coding in Python online as you go through the chapters.

Notes

1 It's said that fans were so infatuated with Liszt's piano "rockstar" status that they fought over his silk handkerchiefs and velvet gloves at his performances.
2 We recommend https://www.w3schools.com/python/.

Interlude 1
BASIC POP BEAT

In this interlude we're going to get familiar with the TunePad interface by creating a basic rock beat in the style of songs like *Roses* by SAINt JHN. You can follow along online by visiting
https://tunepad.com/interlude/pop-beat

STEP 1: Deep listening

It's good to get in the habit of deep listening. Deep listening is the practice of trying every possible way of listening to sounds. Start by loading a favorite song in a streaming service and listening—really listening—to it. Take the song apart element by element. What sounds do you hear? How are they layered together? When do different parts come into the track and how do they change over time? Think about how the producer balances sounds across the frequency spectrum or opens up space for transitions in the lyrics. Try focusing on just the drums. Can you start to recognize the individual percussion sounds and their rhythmic patterns?

STEP 2: Create a new TunePad project

Visit https://tunepad.com on a laptop or Chromebook and set up an account.[1] If you already have a gmail address, you can use your existing account. After signing in, click on the **New Project** button to create an empty project workspace. Your project will look something like this (Figure 1.3):

DOI: 10.4324/9781003033240-2

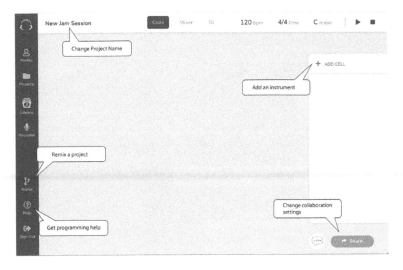

Figure 1.3 TunePad project workspace.

STEP 3: Kick drums

In your project window, click on the **ADD CELL** button and then select **Drums** (Figure 1.4).

In TunePad you can think of a "cell" as an instrument that you can program to play music. Name the new instrument "Kicks" and then add this Python code.

```
1  # play four kick drums
2  playNote(1)
3  playNote(1)
4  playNote(1)
5  playNote(1)
```

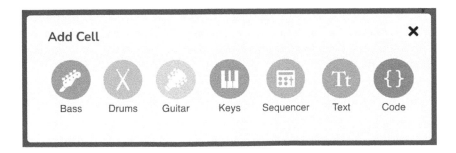

Figure 1.4 Selecting instruments in TunePad.

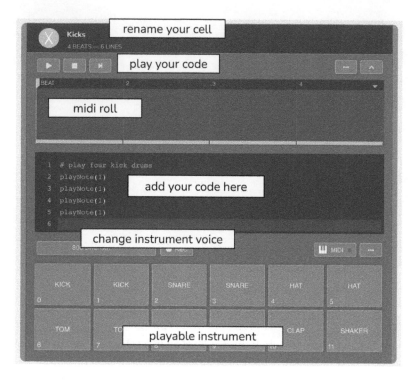

Figure 1.5 Parts of a TunePad cell.

When you're done, your project should look something like Figure 1.5.

Go ahead and press the Play button at the top left to hear how this sounds. Congratulations! You've just written a Python program.

Syntax errors

Occasionally your code won't work right and you'll get a red error message box that looks something like Figure 1.6. This kind of error message is called a "syntax" error. In this case, the code was written as **playnote** with a lowercase "n" instead of an uppercase "N". You can fix this error by changing the code to read **playNote** (with an uppercase "N") on line 2 (Figure 1.6).

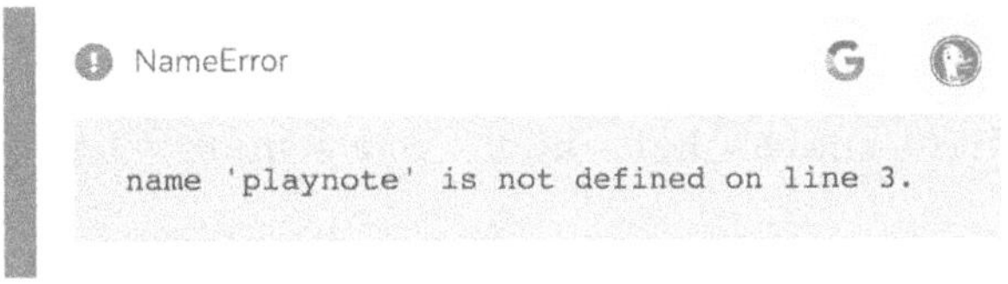

name 'playnote' is not defined on line 3.

Figure 1.6 Python syntax error in TunePad.

STEP 4: Snare drums

In your project window, click on the **ADD CELL** button again and select **Drums**. Now you should have two drum cells one appearing above the other in your project. Name the second instrument "Snare Drums" and then add this Python code.

```
1  # play two snare drums on the up beats only
2  rest(1)          # skip a beat
3  playNote(2)      # play a snare drum sound
4  rest(1)
5  playNote(2)
```

You might start to notice the text that comes after the hashtag symbol (#) is a special part of your program. This text is called a *comment*, and it's for human coders to help organize and document their code. Anything that comes after the hashtag on a line is ignored by Python. Try playing this snare drum cell to hear how it sounds. You can also play the kick drum cell at the same time to see how they sound together.

STEP 5: Hi-hats

Click on the **ADD CELL** button again to add a third drum cell. Change the title of this cell to be "Hats" and add the following code:

```
1  # play four hats between the kicks and snares
2  rest(0.5)                    # rest for half a beat
3  playNote(4, beats=0.5)    # play a hat for half a beat
4  rest(0.5)
5  playNote(4, beats=0.5)
6  rest(0.5)
7  playNote(4, beats=0.5)
8  rest(0.5)
9  playNote(4, beats=0.5)
```

When you play all three of the drum cells together, you should hear a basic rock beat pattern:

kick – hat – snare – hat – kick – hat – snare – hat

STEP 6: Fix your kicks

You might notice that the kick drums feel a little heavy in this mix. We can make some space in the pattern by resting on the up beats (beats 2 and 4) when the snare drums are playing. Scroll back up to your **Kick drum cell** and change the code to look like this:

```
1    # play kicks on the down beats only
2    playNote(1)
3    rest(1)
4    playNote(1)
5    rest(1)
6    playNote(1)
7    rest(1)
8    playNote(1)
9    rest(0.5)                 # rest a half beat
10   playNote(1, beats = 0.5)  # half beat pickup kick
```

STEP 7: Adding a bass line

Add a new cell to your project, but this time select **Bass** instead of **Drums.** Once the cell is loaded up, change the voice to **Plucked Bass** (Figure 1.7):

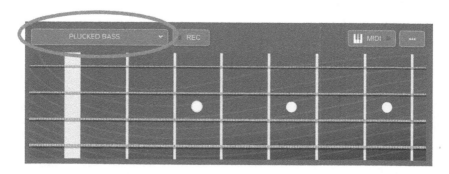

Figure 1.7 Selecting an instrument's voice in TunePad.

Entering this code to create a simplified bass line in the style of *Roses* by SAINt JHN. When you're done, try playing everything together to get the full sound.

```
1    playNote(5, beats=0.5)  # start on low F
2    playNote(17, beats=0.5) # up an octave
3    rest(1)
4
5    playNote(10, beats=0.5) # A sharp
6    playNote(22, beats=0.5) # up an octave
7    rest(1)
8
9    playNote(8, beats=0.5)  # G sharp
10   playNote(20, beats=0.5) # up an octave
11   rest(0.5)
12
13   playNote(8, beats=0.5)   # G sharp - G - G
14   playNote(12, beats=0.5)
15   playNote(24, beats=0.5)
16
17   playNote(10, beats=0.75) # C sharp
19   playNote(22, beats=0.25) # D sharp
```

Note

1 We recommend using the free Google Chrome browser for the best overall experience.

2 Rhythm and tempo

This chapter dives into the fundamentals of **rhythm** in music. We start with the beat—what it is, how it's measured, and how we can visualize the beat to compose, edit, and play music. From there we'll provide examples of some common rhythmic motifs from different genres of music and how to code them with Python. The main programming concepts for this chapter include loops, variables, calling function, and passing parameter values. This chapter covers a lot of ground, but it will give you a solid start on making music with code.

2.1 Beat and tempo

The **beat** is the foundation of rhythm in music. The term *beat* has a number of different meanings in music,[1] but this chapter uses it to mean a unit of time, or how long an individual note is played—for example, "rest for two beats" or "play a note for half a beat". Based on the beat, musical notes are combined in repeated patterns that move through time to make rhythmic sense to our ears.

Tempo refers to the speed at which the rhythm moves, or how quickly one beat follows another in a piece of music. As a listener, you can feel the tempo by tapping your foot to the rhythmic pulse. The standard way to measure tempo is in beats per minute (**BPM** or **bpm**), meaning the total number of beats played in one minute's time. This is almost always a whole number like 60, 120, or 78. At a tempo of 60 bpm, your foot taps 60 times each minute (or one beat per second). At 120 bpm, you get 2 beats

DOI: 10.4324/9781003033240-3

every second; and, at 90 bpm, you get 1.5 beats every second. Later in this chapter when you start working with TunePad, you can set the tempo by clicking on the bpm indicator in the top bar of a project (see Figure 2.1).

Different genres of music have their own typical tempo ranges (although every song and every artist is different). For example, hip-hop usually falls in the 60–110 bpm range, while rock is faster in the 100–140 bpm range. House/techno/trance is faster still, with tempos between 120 and 140 bpm.

Genre	Tempo Range (BPM)
Rock	100–140
R&B	60–80
Pop	100–132
Reggae	60–92
Hip-hop	60–110
Dubstep	130–144
Techno	120–140
Salsa	140–250
Bachata	120–140

It takes practice for musicians to perform at a steady tempo, and they sometimes use a device called a **metronome** to help keep their playing constant with the pulse of the music. You can create a simple metronome in TunePad using four lines of code in a drum cell. This works best if you switch the instrument to **Drums → Percussion Sounds**.

```
playNote(3, velocity = 100) # louder 1st note
playNote(3, velocity = 60)
playNote(3, velocity = 60)
playNote(3, velocity = 60)
```

You can adjust the tempo of your metronome with the bpm indicator (Figure 2.1). As this example illustrates, computers excel at keeping a perfectly steady tempo. This is great if you want precision, but there's also a risk that the resulting music will sound too rigid and machine-like. When real people play music they often speed up or slow down, either for dramatic effect or just as a result of being a human. Depending on the genre, performers might add slight variations in rhythm called swing or shuffle,

Figure 2.1 TunePad project information bar. You can click on the tempo, time signature, or key to change the settings for your project.

that's a kind of back and forth rocking of the beat that you can feel almost more than you can hear. We'll show you how to add a more human feel to computer generated music later in the book.

2.2 Rhythmic notations

Over the centuries, musicians and composers have developed many different written systems to record and share music. With the invention of digital production software, a number of other interactive representations for mixing and editing have become common as well. Here are four common visual representations of the same rhythmic pattern. The pattern has a total duration of four beats and can be counted as "1 and 2, 3 and 4". The first two notes are ½ beats long followed by a note that is 1 beat long. Then the pattern repeats.

2.2.1 Representation 1: Standard Western music notation

The first representation (below) shows standard music notation (or Western notation), a system of recording notes that has been developed over many hundreds of years. The two thick vertical lines on the left side of the illustration indicate that this is rhythmic notation, meaning that there is no information about musical pitch, only rhythmic timing. The dots on the long horizontal lines are notes whose shapes indicate the duration of each sound to be played. Sometimes different percussion instruments will have their notes drawn on different lines. We'll describe what the various note symbols mean in more detail in Figure 2.2.

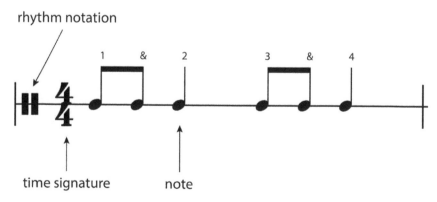

Figure 2.2 Standard notation example.

Figure 2.3 Waveform representation of Figure 2.2.

2.2.2 Representation 2: Audio waveforms

The second representation shows a visualization of the actual audio wave-form that gets sent to the speakers when you play music. The waveform shows the amplitude (or volume) of the audio signal over time. The next chapter talks more about audio waveforms, but for now you can think of a waveform as a graph that shows the literal intensity of the vibration of your speakers over time. When you compose a beat in TunePad, you can switch to the waveform view by clicking on the small dropdown arrow at the top-left side of the timeline (Figure 2.3).

2.2.3 Representation 3: Piano (MIDI) roll

The third representation shows a piano roll (or MIDI (Musical Instrument Digital Interface) roll). This uses solid lines to show individual notes. The length of the lines represents the length of individual notes, and the vertical position of the lines represents the percussion sound being played (kick drums and snare drums in this case). This representation is increasingly common in music production software. Many tools even allow for drag and drop interaction with the individual notes to compose and edit music (Figure 2.4).

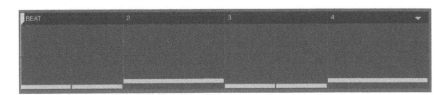

Figure 2.4 Piano or MIDI roll representation of Figure 2.2.

2.2.4 Representation 4: Python code

A final representation for now shows Python code in TunePad. In this representation, the duration of each note is set using the **beats** parameter of the **playNote** function calls.

```
playNote(2, beats = 0.5)
playNote(2, beats = 0.5)
playNote(6, beats = 1)

playNote(2, beats = 0.5)
playNote(2, beats = 0.5)
playNote(6, beats = 1)
```

Each of these representations has advantages and disadvantages; they are good for conveying some kinds of information and less good at conveying others. For example, standard rhythm notation has been refined over centuries and is accessible to an enormous, worldwide community of musicians. On the other hand, it can be confusing for people who haven't learned how to read sheet music. The timing of individual notes is communicated using tails and flags attached to the notes, but there's no consistent mapping between horizontal space and timing.

The audio waveform is good at showing what the sound *actually* looks like—how long each note rings out ("release") and how sharp its onset is ("attack"). It's helpful for music production, mixing, and mastering. On the other hand, waveforms don't really tell you much about the pitch of a note or its intended timing as recorded by the composer.

The Python code is easier for computers to read than humans—it's definitely not something you would hand to a musician to sight read. On the other hand, it has the advantage that it can be incorporated into computer *algorithms* and manipulated and transformed in endless ways.

There are many, many other notation systems designed to transcribe a musical performance—what we hear at a live performance—onto a sheet of paper or a computer screen. Each of these representations was invented for a specific purpose and/or genre of music. You might pick a representation based on the context and whether you're in the role of a musician (and what kind of instrument you play), a singer, a composer, a sound engineer, or a producer. Music notation systems are as rich and varied as the cultures and musical traditions that invented them. One nice thing about working with software is that it's easy to switch between multiple representations of music depending on the task we're trying to accomplish.

Whole Note (4 beats) ——— playNote(60, beats = 4)

Half Notes (2 beats) ——— playNote(60, beats = 2)

Quarter Notes (1 beat) ——— playNote(60, beats = 1)

Eighth Notes (1/2 beat) ——— playNote(60, beats = 0.5)

Sixteenth Notes (1/4 beat) ——— playNote(60, beats = 0.25)

Figure 2.5 Common note symbols starting with a whole note (four beats) on the top down to 16th notes on the bottom. The notes on each new row are half the length of the row above.

2.3 Standard rhythmic notation

This section will review a standard musical notation system that has roots in European musical traditions. This system is versatile and has been refined and adapted over a long period of time across many countries and continents to work with an increasingly diverse range of instruments and musical genres. We're starting with percussive rhythmic note values in this chapter, and we'll move on to working with pitched instruments in Chapter 3.

Figure 2.5 shows the most common symbols used in rhythmic music notation. Notes are represented with oval-shaped dots that are either open or closed. All notes except for the whole note (top) have tails attached to them that can point either up or down. It doesn't matter which direction (up or down) the tail points. Notes that are faster than a quarter note also have horizontal flags or beams connected to the tails. Each additional flag or beam indicates that the note is twice as fast.

Symbol	Name	Beats	TunePad code
O	**Whole Note** Larger open circle with no tail and no flag.	4	playNote(1, beats = 4)
𝅗𝅥	**Half Note** Open circle with a tail and no flag.	2	playNote(1, beats = 2)
𝅘𝅥	**Quarter Note** Solid circle with a tail and no flag.	1	playNote(1, beats = 1)

♪	**Eighth Note** Solid circle with a tail and one flag or bar.	0.5 or ½	`playNote(1, beats = 0.5)`
♪	**Sixteenth Note** Solid circle with a tail and two flags or bars.	0.25 or ¼	`playNote(1, beats = 0.25)`
♩.	**Dotted Half Note** Open circle with a tail. The dot adds an extra beat to the half note.	3	`playNote(1, beats = 3)`
♩.	**Dotted Quarter Note** Solid circle with a tail. The dot adds an extra half-beat.	1.5	`playNote(1, beats = 1.5)`
♪.	**Dotted Eighth Note** Solid circle with tail and one flag. The dot adds an extra quarter beat.	0.75	`playNote(1, beats = 0.75)`

Standard notation also includes *dotted notes*, where a small dot follows the note symbol. With a dotted note, you take the original note's duration and add half of its value to it. So, a dotted quarter note is 1.5 beats long, a dotted half note is 3 beats long, and so on.

There are also symbols representing different durations of silence or "rests".

Symbol	Name	Beats	TunePad code
▬	**Whole Rest**	4	`rest(beats = 4)`
▬	**Half Rest**	2	`rest(beats = 2)`
𝄽	**Quarter Rest**	1	`rest(beats = 1)`
𝄾	**Eighth Rest**	0.5 or ½	`rest(beats = 0.5)`
𝄿	**Sixteenth Rest**	0.25 or ¼	`rest(beats = 0.25)`

2.4 Time signatures

In standard notation, notes are grouped into segments called **measures** (or bars). Each measure contains a fixed number of beats, and the duration of all the notes in a measure should add up to that amount. The relationship between measures and beats is represented by a fraction called a **time signature**. The numerator (or top number) indicates the number of beats in the measure, and the denominator (bottom number) indicates the beat duration.

4/4	**Four-Four Time or "Common Time"** There are 4 beats in each measure, and each beat is a quarter note. This time signature is sometimes indicated using a special symbol:	**C**	♩♩♩♩
2/2	**Two-Two Time or "Cut Time"** There are 2 beats in each measure, and the beat value is a half note. Cut time is sometimes indicated with a 'C' with a line through it.	**₵**	𝅗𝅥 𝅗𝅥
2/4	**Two-Four Time** There are 2 beats in each measure, and the quarter note gets the beat.		♩ ♩
3/4	**Three-Four Time** There are 3 beats in each measure, and the quarter note gets the beat.		♩♩♩
3/8	**Three-Eight Time** There are 3 beats in each measure, and the eighth note gets the beat.		♪ ♪ ♪

The most common time signature is 4/4. It's so common, in fact, that it's referred to as **common time**. It's often denoted by a C symbol shown in the table above. In common time, there are four beats to each measure, and the quarter note "gets the beat" meaning that one beat is the same as one quarter note.

Vertical lines separate the measures in standard notation. In the example below, there are two measures in 4/4 time (four beats in each measure, and each beat is a quarter note).

If you have a time signature of 3/4, then there are three beats per measure, and each beat's duration is a quarter note. Some examples of songs in 3/4 time are *My Favorite Things* from *The Sound of Music*, *My 1st Song* by Jay Z, *Manic Depression* by Jimi Hendrix, and *Kiss from a Rose* by Seal.

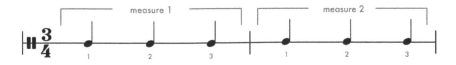

If those notes were eighth notes, it would look like this:

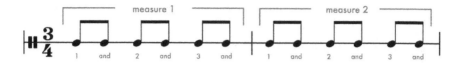

Other common time signatures include 2/4 time (with two quarter note beats per measure) and 2/2 time (with two *half note* beats in each measure). With 2/2 there are actually four quarter notes in each measure because one half note has the same duration as two quarter notes. For this reason, 2/2 time is performed similarly to common time, but is generally faster. It is referred to as **cut time** and is denoted by a C symbol with a line through it (see table above).

You can adjust the time signature of your TunePad project by clicking on the time indicator in the top bar (see Figure 2.1).

2.5 Percussion sounds and instruments

Working with rhythm, you'll come across lots of terminology for different percussion instruments and sounds. Here's a quick rundown on some of the most common drum sounds that you'll work with in digital music (Figure 2.6).

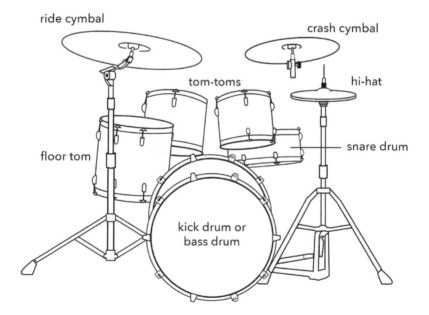

Figure 2.6 Drums in a typical drum kit.

Figure 2.6 was modified from an original drawing by Syed Wamiq Ahmed Hashmi (commons.wikimedia.org/ wiki/User:Syed_Wamiq_Ahmed_Hashmi). Creative Commons License creativecommons.org/licenses/by-sa/3.0.

Drum names	Description	TunePad note number
Kick or bass drum	The kick drum (or bass drum) makes a loud, low thumping sound. Kicks are commonly placed on beats 1 and 3 in rock, pop, house, and electronic dance music. In other genres like hip-hop and funk, kick drums are very prominent, but their placement is more varied.	0 and 1
Snare	Snare drums make a recognizable sharp staccato sound that cuts across the frequency spectrum. They are built with special wires called snares that give the drum its unique snapping sound. Snare drums are commonly used on beats 2 and 4.	2 and 3
Hi-hat	The hi-hat is a combination of two cymbals sandwiched together on a metal rod. A foot pedal opens or closes the cymbals together. In the closed position the hi-hat makes a bright tapping sound. In the open position the cymbal is allowed to ring out. Hi-hats have become an integral part of rhythm across almost all genres of popular music.	4 (closed) 5 (open)
Low, mid, high tom	Tom drums (tom-toms) are cylindrical drums that have a less snappy sound than the snare drum. Drum kits typically have multiple tom drums with slightly different pitches (such as low, mid, and high).	6, 7, 8
Crash cymbal	A large cymbal that makes a loud crash sound, often used as a percussion accent.	9
Claps and shakers	Different TunePad drum kits include a range of other percussion sounds common in popular music including various claps, shakers, and other sounds.	10 and 11

2.5.1 808 Drum kit

Released in the early 1980s, the Roland 808 drum machine was a hugely influential sound in early hip-hop music (and other genres as well). The 808 used electronic synthesis techniques to create synthetic replicas of drum sounds like kicks, snares, hats, toms, cowbells, and rim shots. Tinkerers would also open up the 808s and hack the circuits to create entirely new sounds. Today 808s usually refers to low, booming bass lines that were first generated using tweaked versions of the 808s' kick drums. TunePad's default drum kit uses samples that sound like the original electronically synthesized 808s (Figure 2.7).

2.5.2 Selecting TunePad instruments

When you're coding in TunePad, the sound that your code makes will depend on the instrument you have selected. If you're coding a rhythm,

Figure 2.7 Roland 808 drum sequencer.
The photograph shown in Figure 2.7 and at the beginning of Interlude 10 is by Brandon Daniel (flickr.com/photos/54581307@N00). Creative Commons License creativecommons.org/licenses/by-sa/2.0.

you can choose from several different drum kits including an 808 and rock kits. You can change the instrument by clicking on the selector shown below (Figure 2.8).

2.6 Coding rhythm in Python

Let's start coding! Before beginning to code, there are a few quick things that you should keep in mind.

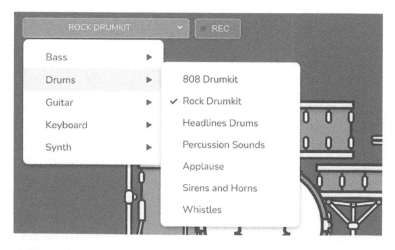

Figure 2.8 Changing an instrument's voice in TunePad.

2.6.1 *Syntax errors*

Python is a text-based language, which means that you're going to be typing code that has to follow strict grammatical rules. When you speak a natural language like English, grammar is important, but you can usually bend or break the rules and still get your message across. When you say something ambiguous it can be ironic, humorous, or poetic. This isn't the case in Python. Python has no sense of humor and no appreciation for poetry. If you make a grammatical mistake in coding, Python gives you a message called a **syntax error**. These messages can be confusing, but they're there to help you fix your code in the same way that a spell checker helps you fix typos. Here's what a syntax error looks like in TunePad (Figure 2.9).

This line of code was missing a parenthesis symbol, which generated the error message "**bad input on line 5**". Notice that Python is giving you hints about where the problems are and how to fix them, but those hints aren't always that helpful and can be frustrating for beginners. We'll give you practice fixing syntax later in this chapter.

2.6.2 *Flow of control*

A Python program is made up of a list of statements. For the most part, each statement goes on its own line in your program. Python will read and perform each line of code from the top to the bottom in the order that you write them. In programming this is called the **flow of control**. This is similar to the way you would read words in a book or notes on a line of sheet music. The difference is that programming languages also have special rules that let you change the flow of control. Those rules include **loops** (which repeat some part of your code multiple times), **conditional logic** (which runs some part of your code only if some condition is met), and **user-defined functions** (which lets you create your own functions that can be called). We'll talk about these special "**control structures**" later in the book.

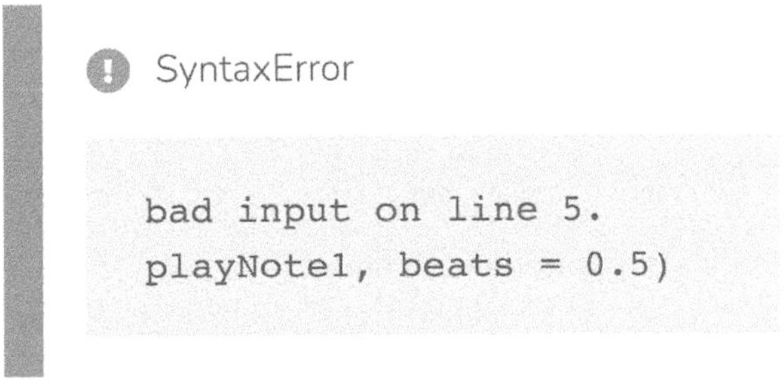

! SyntaxError

```
bad input on line 5.
playNote1, beats = 0.5)
```

Figure 2.9 Example of a Python syntax error in TunePad. This line of code was missing a parenthesis symbol.

2.7 Calling functions

Almost everything you do in Python involves *calling* functions. A function (sometimes called a command or an instruction) tells Python to do something or to compute a value. For example, the **playNote** function tells TunePad to make a sound. There are three things you have to do to call a function:

First, you have to write the name of the function. Functions have one-word names (with no spaces) that can consist of letters, numbers, and the underscore _ character. Multi-word functions will either use the underscore character between words as in:

```
my_multi_word_function()
```

or each new word will be capitalized as in **playNote()**.

Second, after you type the name of the function, you have to include parentheses. The parentheses tell Python that you're calling a function.

Third, you include any *parameters* that you want to *pass* to the function in between the left and right parentheses. A parameter provides extra information or tells the function how to behave. For example, the **playNote** statement needs at least one parameter to tell it which note or sound to play. Sometimes functions accept multiple parameters (some of which can be optional). The **playNote** function accepts several optional parameters described in the next section. Each additional parameter is separated with a comma (Figure 2.10).

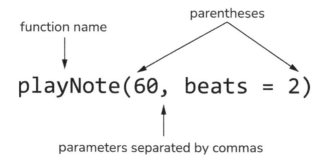

Figure 2.10 Calling the playNote function in TunePad with two parameters inside the parentheses.

2.8 The playNote function

The **playNote** function tells TunePad to play a percussion sound or a musical note. The **playNote** function accepts up to four parameters contained within the parentheses.

```
playNote(1, beats = 1, velocity = 100, sustain = 0)
```

Name	Description
note	This is a **required** parameter that says which note or percussion sound to play. The kind of sound depends on which instrument you have selected in TunePad for this code. You can play more than one note at the same time by enclosing notes in square brackets.
beats	An **optional** parameter that says how long to play the note. The TunePad *playhead* will be moved forward by the duration given. This parameter can be a whole number (like 1 or 2), a decimal number (like 1.5 or 0.25), or a fraction (like 1/2).
velocity	An **optional** parameter that says how loud to play the note or sound. A value of 100 is full volume, and a value of 0 is no volume (muted). Velocity is a technical term in digital music that means how fast or how hard you hit the instrument. You might imagine it as how loud a drum sounds based on how hard it gets hit.
sustain	An **optional** parameter that allows a note to ring out for an additional number of beats without advancing the playhead.

2.8.1 *Optional parameters*

Sometimes parameters are *optional*, meaning that they have a value that gets provided by default if you don't specify one. For **playNote**, only the note parameter is required. If you don't pass the other parameters, it provides values for you by default. You can also include the *names* of parameters in a function call. For example, all four of the lines below do exactly the same thing; they play a note for one beat. The first two use parameters without their names. The second two include the names of the parameter, followed by the equals sign (=), followed by the parameter value.

```
playNote(60)                  # the beats parameter is optional
playNote(60, 1)               # with the beats parameter set to 1
playNote(60, beats = 1)       # with a parameter name for beats
playNote(note = 60, beats = 1) # with a parameter name for note and beats
```

2.8.2 Comments

In the code above, some of the text appears after hashtag (#) symbols on each line. This text is called a **comment**. A comment is a freeform note that programmers add to make their code easier to understand. Comment text is ignored by Python, so you can write anything you want after the hashtag symbol on a line. You can also use a hashtag at the beginning of a line to temporarily disable code. This is called "commenting out" code.

2.9 The rest function

Silence is an important element of music. The **rest** function generates silence, or a break in the sound. It only takes one parameter, which is the length of time the rest is held. So, **rest(beats = 2)** will trigger a rest for a length of two beats. If you don't provide a parameter, **rest** uses a value of 1.0 by default.

```
rest()                  # rest for one beat
rest(1.0)               # rest for one beat
rest(0.25)              # rest for one quarter beat
rest(beats = 0.25)      # rest for one quarter beat
```

2.10 Examples of playNote and rest

Let's try a few examples of **playNote** and **rest** to get warmed up. This rhythm plays two eight notes (beats = 0.5) followed by a quarter note (beats = 1). The pattern then repeats a second time.

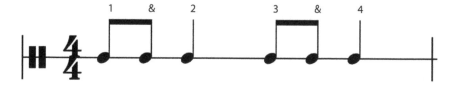

Here's how we would code this in TunePad with a kick drum and snare:

```
playNote(1, beats = 0.5)    # play a kick drum (1) for half a beat
playNote(1, beats = 0.5)
playNote(2, beats = 1)      # play snare (2) for one beat
playNote(1, beats = 0.5)    # play kick (1) for half a beat
playNote(1, beats = 0.5)
playNote(2, beats = 1)      # play snare (2) for one beat
```

Here's another example that plays a quarter note followed by a rest of 0.5 beats followed by an eight note (beats = 0.5). The pattern is repeated two times in a row:

```
playNote(2, beats = 1)      # play a snare drum (2) for one beat
rest(beats = 0.5)           # rest for half a beat
playNote(1, beats = 0.5)    # play a kick drum (1) for half a beat
playNote(2, beats = 1)      # play a snare drum (2) for one beat
rest(beats = 0.5)           # rest for half a beat
playNote(1, beats = 0.5)
```

Here's a third example that plays eight notes in a row, each an eight note (beats = 0.5). See if you can write the code to make this pattern.

```
# write your Python code here
```

2.11 Loops

All of the examples in the previous section included repeated elements. And, if you listen closely, you can hear repeated elements at all levels of music. There are repeated rhythmic patterns, recurring melodic motifs, and storylines defined by song sections that get repeated and elaborated. It turns out that there are many circumstances in both music and computer programming where we want to repeat something over and over again.

To show how we can take advantage of some of the capabilities of Python, let's start with the last example from the previous section where we wanted to tap out a run of eighth notes (0.5 beats) on the hi-hat. One way to program that rhythm would be to just type in eight **play-Notes** in a row (below left).

```
1   playNote(4, beats=0.5)
2   playNote(4, beats=0.5)
3   playNote(4, beats=0.5)
4   playNote(4, beats=0.5)
5   playNote(4, beats=0.5)
6   playNote(4, beats=0.5)
7   playNote(4, beats=0.5)
8   playNote(4, beats=0.5)
```

⇨

```
1   for i in range(8):
2       playNote(4, beats = 0.5)
3       print(i)
```

This will get the job done, but there are a few problems with this style of code. One problem is that it violates one of the most important character traits of a computer programmer—laziness! A lazy programmer is someone who works smart, not hard. A lazy programmer avoids doing repetitive, error-prone work. And, a lazy programmer knows that there are some things that the computer can do better than a human can.

In Python (and just about any other programming language), when we want to do something multiple times, we can use a loop. Python has a number of different kinds of loops, but, in this case, our best option is something called a **for** loop. The version of the code on the right repeats eight times in a row. For each iteration of the loop, the TunePad **playNote** function gets called.

With the original code on the left, we had to do a lot of typing (or, more likely, copying and pasting) to enter our program—a warning sign that we're not being lazy enough. We generated a lot of repetitive code, which

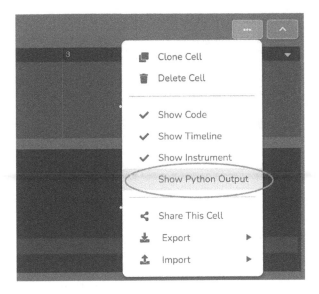

Figure 2.11 How to show the print output of your code in a TunePad cell.

makes the program harder to read (not as legible), error prone, and not as elegant as it could be. The right-side code above accomplishes the same thing with just three lines of code instead of eight.

Finally, the code on the left is harder to change and reuse. What if we wanted to use a different drum sound (like a snare instead of a hat)? Or, what if we wanted to tap out a run of 16 sixteenth notes instead of 8 eighth notes? We would have to go through the code line by line making the same change over and over again. This is a slow, error-prone process that is definitely not lazy or elegant.

To see why this is better, try changing the code on the right so that it plays 16 sixteenth notes instead of 8 eighth notes. Or try changing the drum sound from a hat to something else. The **print** statement on line 3 is just there to help you see what's going on with your code. If you click on the **Show Python Output** option, you can see how the variable called *i* (that gets created on line 1) counts up from 0 to 7 (Figure 2.11).

Here's more detail about the anatomy of a for loop (Figure 2.12):

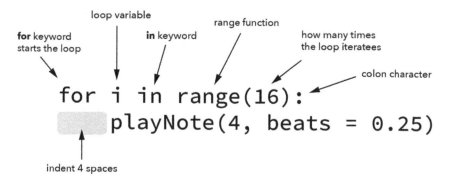

Figure 2.12 Anatomy of a for loop in Python.

A for loop with the range function:

- begins with the **for** keyword
- includes a loop **variable** name; this can be anything you want (above it is "**i**"). Each time the loop goes around, the loop variable is incremented by one.
- includes the **in** keyword
- includes the **range** function that says how many times to repeat (above, this is 16)
- includes a colon **:**
- includes a block of code **indented** by 4 spaces

Python uses **indentation** to determine what's *inside* the loop, meaning it's the section of code that gets repeated multiple times. The indented block of code is repeated the total number of times specified by **range**. Let's try adding a few extras to the previous example. In the version below, we add a run of sixteenth notes for the last beat.

```
1  for i in range(6):
2      playNote(4, beats = 0.5)
3
4  for i in range(4):
5      playNote(4, beats = 0.25)
```

But there are lots of other things we could do as well. If we wanted to play an even faster run, we could use code like:

```
for i in range(8):
    playNote(4, beats = 0.125)
```

Or, if we wanted to play a triplet that divides a half-beat into three equal parts, we could do something like this:

```
for i in range(3):
    playNote(4, beats = 0.25 / 3) # divide into 3 parts
```

If you open this example in TunePad, you can experiment with different combinations of numbers to get different effects:
https://tunepad.com/examples/loops-and-hats

2.12 Variables

A **variable** is a name you give to some piece of information in a Python program. You can think of a variable as a kind of nickname or alias. Similar to loops, variables help make your code more elegant, easier to read, and easier to change in the future. For example, the code on the left plays a drum pattern without variables, and the code on the right plays the same thing with variables. Notice how the variables help make the code easier to understand because they give us descriptive names for the various drum sounds instead of just numbers.

1	`playNote(0)`
2	`playNote(4)`
3	`playNote(2)`
4	`playNote(4)`

⇨

1	`kick = 0`
2	`hat = 4`
3	`snare = 2`
4	
5	`playNote(kick)`
6	`playNote(hat)`
7	`playNote(snare)`
8	`playNote(hat)`

In the version on the right we defined a variable called **kick** on line 1, a variable called **hat** on line 2, and a variable called **snare** on line 3. Each variable is *initialized* to a different number for the corresponding drum sound. It's also possible to change the value of a variable later in the program by assigning it a different number.

```
kick = 0
playNote(kick)  # plays sound 0
kick = 1        # set kick to a different value
playNote(kick)  # plays sound 1
```

Variable names can be anything you want as long as they're one word long (no spaces) and consist only of letters, numbers, and the underscore character (_). Variable names cannot start with a number, and they can't be the same as any existing Python keyword.

As you begin to get comfortable with code and to exercise your creativity, you'll find yourself wanting to experiment with sounds. You might want to try different sounds for the same rhythmic pattern, maybe change a high-hat sound to a shaker to get a more organic feel. Using variables makes it easy to experiment by changing values around.

Here's another example with a hi-hat pattern. Imagine that you really like this pattern, but you're wondering how it would sound with a different percussion instrument. Maybe you want to change the 4 sound to a shaker sound (like 11). The nice thing about variables is that you can give them just about any name you want as long as Python is not already using that name for something else. This way you can make the name meaningful to you. So, for our shaker example we could create a variable with a meaningful name like **shake** and set it equal to 11. When you use the variable **shake** you are inserting whatever number is currently assigned to it.

```
for i in range(8):
    playNote(4,
    beats = 0.5)
for i in range(8):
    playNote(4,
    beats = 0.25)
for i in range(4):
    playNote(4,
    beats = 0.5)
```

⇨

```
shake = 11
for i in range(8):
    playNote(shake,
    beats = 0.5)
for i in range(8):
    playNote(shake,
    beats = 0.25)
for i in range(4):
    playNote(shake,
    beats = 0.5)
```

As you progress with coding, you'll find that loops and variables help create a smoother workflow that gives you more flexibility, freedom, and creative power.

Try out using variables with the following exercise: https://tunepad.com/examples/variables

2.13 More on syntax errors

We've mentioned that Python code is like a language with strict grammatical rules called syntax. When you make a mistake in coding—and everyone makes coding mistakes all the time—Python will give you feedback about what the error is and approximately what line it's on. For instance, if you've been trying to code the exercises in this chapter, you may have seen a message like Figure 2.13.

This is telling us that there is an error on line 6 that can be fixed by changing the text, "ployNote". When you are using a variable or function in your code, Python is expecting you to type it *exactly* the same as it was defined. A simple typo can stop your program from running, but it's also easily fixed. Here we just need to update the line to say **playNote** instead of **ployNote**.

Other syntax errors are trickier. The message in Figure 2.14 is confusing because the problem is actually on line 1 even though the syntax error says line 2. The problem is a missing right parenthesis on line 1 (Figure 2.14).

> ⓘ NameError
>
> ```
> name 'ployNote' is not defined on line 6.
> ```

Figure 2.13 Example of a Python syntax error. The command 'ployNote' should instead say 'playNote'.

> ⓘ SyntaxError
>
> ```
> bad input on line 2.
> rest(1)
> ```

Figure 2.14 Example of a Python syntax error. Here the problem is actually on line 1, not line 2.

```
1   playNote(60
2   rest(1)
```

One technique coders use to find the source of errors like this is to comment out lines of code before and after an error. For example, to comment out the first line of the code above, we could change it to look like this:

```
# playNote(60
rest(1)
```

Adding the hashtag at the beginning of the first line means that Python ignores it, in this case fixing the syntax error and giving us another clue about the source of the problem.

Another surprisingly helpful trick is to just paste your error message verbatim into your favorite search engine. There are huge communities of Python coders out there who have figured out how to solve almost every problem with code imaginable. You can often find a quick fix to your problem just by browsing through a few of the top search results.

If you want practice fixing syntax errors in your code, you can try one of our mystery-melody challenges on TunePad:

https://tunepad.com/examples/mystery-melody.

2.14 The playhead

The timing of notes in TunePad is determined by the position of an object called the **playhead**. In the early days of music production, recordings were made using analog tape. Sound wave signals coming from a microphone or some other source were physically stored on magnetic tape using a mechanism called a **record head**. As the tape moved by, the record head would inscribe patterns of magnetic material inside of the tape, thus creating a recording of the music. To play the recording back, a **playhead** would pick up fluctuations in the tape's magnetic material and convert it back into sound waves for listeners to hear. Fast forward to the digital realm. We no longer have playheads or record heads, but we maintain that metaphor when referring to the notion of sound moving in time. The concept of a playhead is common across audio production software as the point in time where audio is playing.

In TunePad, when you place a note with the **playNote** function, it advances the playhead forward in time by the duration of the note specified by the **beats** parameter. There are several functions available to get information about the position of the playhead and move it forward or backward in time.

Function	Description
getPlayhead()	Returns the current position of the playhead in beats. Note that **getPlayhead** returns the elapsed number of beats, which means that if the playhead is at the beginning of a track the function will return 0. If 1.5 beats have elapsed, **getPlayhead** will return 1.5. If 40 beats have elapsed, it will return 40, and so on.
getMeasure()	Returns the current measure as an integer value. Note that **getMeasure** returns an elapsed number of measures. So, if the playhead is at the beginning of the track or anywhere before the end of the first measure, the function will return 0. If the playhead is greater than or equal to the start of the second measure, **getMeasure** will return 1, and so on.
getBeat()	Returns an elapsed number of beats *within the current measure* as a decimal number. For example, if the playhead has advanced by a quarter beat within a measure, **getBeat** will return 0.25. The value returned by **getBeat** will always be less than the total number of beats in a measure.
fastForward(beats)	Advance the playhead forward by the given number of beats relative to the current position. Note that this is identical to the **rest** function. Negative beat values move the playhead backward.
rewind(beats)	Move the playhead back in time by the given number of beats. This can be a useful way to play multiple notes at the same time. The beats parameter specifies the number of **beats** to move the playhead. Negative values of beats move the playhead forward.

rest(beats)   Advance the playhead forward by the given number
                            of beats without playing a sound. This is identical to
                            the **fastForward** function.

moveTo(beats)   Move the playhead to an arbitrary position. The
                            beats parameter specifies the point that the
                            playhead will be placed as an elapsed number of
                            beats. For example, **moveTo(0)** will move the
                            playhead to the beginning of a track (zero elapsed
                            beats). **moveTo(1)** will place the playhead at the end
                            of the first beat and right before the start of the
                            second beat.

We can control where the playhead is relative to the music we make by
using the **moveTo**, **fastForward**, and **rewind** commands. The **rewind** and
fastForward functions move the playhead backward or forward relative
to the current point in time. The **moveTo** function takes the playhead and
moves it to an arbitrary point in time. In TunePad, the playhead represents
an *elapsed* number of beats. So, to move to the beginning of a track, you
would use **moveTo(0)**, which means zero elapsed beats. To move to the
beginning of the second beat, you would use **moveTo(1)**, which means one
elapsed beat. These commands are useful for adding multiple overlapping
rhythms to a single TunePad cell. We'll see more on how these commands
can be used in Chapter 8.

2.15 Basic drum patterns

Now let's code some foundnational drum patterns There is also a link to
the code in TunePad so that you can play around with the beat and make
it your own.

2.15.1 Four-on-the-floor

The four-on-the-floor pattern is a staple of House, EDM, disco, and pop
music. It has a driving dance beat defined by four kick drum hits on each
beat (thus four beats on the floor). This beat is simple but versatile. You can
spice it up by moving hi-hats around and adding kicks, snares, and other
drums in unexpected places. We can make the basic pattern with just three
instruments: kick, snare, and hats.

	1	·	·	·	2	·	·	·	3	·	·	·	4	·	·	·
hat (4)	●	●	●	●	●	●	●	●	●	●	●	●	●	●	●	●
snare (2)					●								●			
kick (0)	●				●				●				●			

Kick drums are the lowest drum sound in a drum kit. Start by laying down kick drums on each beat of the measure. These low sounds give this pattern a driving rhythmic structure that sounds great at higher tempos. Then we'll add snare hits on the even beats (2 and 4). The snare adds energy and texture to the beat. Finally, we add hi-hats. These are the highest pitch instruments in most drum beats, and they help outline the groove to emphasize the beat.

You can find this example at

https://tunepad.com/examples/four-on-the-floor

```
1    # define instrument variables
2    kick = 0
3    snare = 2
4    hat = 4
5
6    # lay down four kicks (on the floor)
7    playNote(kick)
8    playNote(kick)
9    playNote(kick)
10   playNote(kick)
11
12   moveTo(0)   # reset playhead to the beginning
13
14   # add snares on the even beats
15   rest(1)
16   playNote(snare)
17   rest(1)
18   playNote(snare)
19
20   moveTo(0)   # reset playhead to the beginning
21
22   # hi-hat pattern with a loop!
23   for i in range(8):
24       playNote(hat, 0.5)
```

2.15.2 Blues

Blues is a genre of music that evolved from the African American experience, starting as field songs, evolving into spirituals, and eventually became the Blues. This beat is in 3/4 time, meaning that there are three beats in each measure. You can find this example at

https://tunepad.com/examples/blues-beat

	1	.	.	.	2	.	.	.	3	.	.	.
hat (4)	●		●	●		●	●		●	●		●
snare (2)				●						●		
kick (1)	●		●			●	●		●		●	●

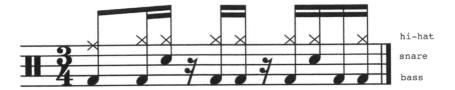

```
1    # define instrument variables
2    kick = 0
3    snare = 2
4    hat = 4
5
6    # lay down kick and snare pattern
7    playNote(kick, beats = 0.5)
8    playNote(kick, beats = 0.25)
9    playNote(snare, beats = 0.25)
10   rest(.25)
11   playNote(kick, beats = 0.25)
12   playNote(kick, beats = 0.25)
13   rest(.25)
14   playNote(kick, beats = 0.25)
15   playNote(snare, beats = 0.25)
16   playNote(kick, beats = 0.25)
17   playNote(kick, beats = 0.25)
18
19   moveTo(0) # reset playhead to beginning
20
21   # add hi-hats
22   playNote (4, beats = 0.25)
23   for i in range(3):
24       rest(0.25)
25       playNote (hat, beats = 0.25)
26       playNote (hat, beats = 0.25)
27   rest(0.25)
28   playNote(hat, beats = 0.25)
```

2.15.3 Latin

Latin beats are known for their syncopated rhythms that emphasize the so-called "weak beats" in a measure. This drum pattern is two measures long. Our kick pattern sounds much like a heartbeat and solidly grounds our entire beat. Our snare is playing a *clave* pattern, which is common in many forms of Afro-Cuban music such as salsa, mambo, reggae, reggaeton, and dancehall. In the second measure, we have hits on the first beat and the second half of the second beat (counts 1 and 2.5). Finally, we add a hi-hat on every eighth note.

	1	.	.	.	2	.	.	.	3	.	.	.	4	.	.	.
hat (4)	●		●		●		●		●		●		●		●	
snare (3)					●				●							
kick (0)	●						●		●						●	

	5	.	.	.	6	.	.	.	7	.	.	.	8	.	.	.
hat (4)	●		●		●		●		●		●		●		●	
snare (3)	●						●						●			
kick (0)	●						●		●						●	

You can find this example at https://tunepad.com/examples/latin-beat.

```
1    # define instrument variables
2    kick = 0
3    snare = 3
4    hat = 4
5
6    # lay down kicks for the heartbeat
7    for i in range(2):
8        playNote(kick, beats = 1.5)
9        playNote(kick, beats = 0.5)
10       playNote(kick, beats = 1.5)
11       playNote(kick, beats = 0.5)
12
13   moveTo(0) # reset playhead
14
15   # add snare
16   rest(1.0)
17   playNote(snare)
18   playNote(snare)
19   rest(1.0)
20   playNote(snare, beats = 1.5)
21   playNote(snare, beats = 1.5)
22   playNote(snare)
23
24   moveTo(0) # reset playhead
25
26   # lay down hi-hats
27   for i in range(16):
28       playNote(hat, beats = 0.5)
```

2.15.4 Reggae

A common reggae beat is the ***one-drop beat***, which gets its name due to the fact there's no hit on the first beat. Rather, the accent is on the third beat which contributes to the strong backbeat and laid back feel in reggae. We're using swung eighth notes for our hi-hats, and adding an open hi-hat hit on the very last note to add texture. The first hit is held for two-thirds of the beat and the second for one-third. Both our kick and snare hit on the third beat. You can find this example at https://tunepad.com/examples/reggae-beat.

	1	.	.	2	.	.	3	.	.	4	.	.
open_hat (5)												●
hat (4)	●		●	●		●	●		●	●		
snare (2)							●					
kick (0)							●					

```
1    # define instrument variables
2    kick = 0
3    snare = 2
4    hat = 4
5    open_hat = 5
6
7    # lay down kick and snare together
8    rest(2)
9    playNote([kick, snare])
10
11   moveTo(0) # reset playhead
12
13   # lay down swung hi-hat pattern
14   for i in range(3):
15       playNote(hat, 2.0 / 3)   # two-thirds
16       playNote(hat, 1.0 / 3)   # one-third
17
18   playNote(hat, 2.0 / 3)
19   playNote(open_hat, 1.0 / 3)
```

2.15.5 Other common patterns

Here are a few other drum patterns in different genres that you can try
coding for yourself.

HIP-HOP (Late–1990s)   90 bpm

	1	.	.	.	2	.	.	.	3	.	.	.	4	.	.	.
open-hat (5)			●								●					
snare (2)					●								●			
kick (1)	●					●			●							

HIP-HOP (Mid-2000s)   85bpm

	1	.	.	.	2	.	.	.	3	.	.	.	4	.	.	.
snare (2)					●								●			
kick (1)	●		●	●			●	●		●	●			●		●

BASIC POP/ROCK   130bpm

	1	.	.	.	2	.	.	.	3	.	.	.	4	.	.	.
snare (2)					●								●			
kick (1)	●								●							

	5	.	.	.	6	.	.	.	7	.	.	.	8	.	.	.
snare (2)					●								●			
kick (1)	●								●							

TRAP (Mid-2010s)   130bpm

	1	.	.	.	2	.	.	.	3	.	.	.	4	.	.	.
hat (4)	●●	●	●	●	●●	●	●	●	●	●	●	●	●●	●	●	●
snare (4)					●								●			
kick (1)	●						●		●							

The double dots mean stuttered hi–hats

POP/HIP-HOP   70 bpm

	1	.	.	.	2	.	.	.	3	.	.	.	4	.	.	.
hat (4)			●				●				●				●	
snare (2)					●								●			
kick (1)	●								●							

WEST COAST BEAT (Late 2010s)   100 bpm

	1	.	.	.	2	.	.	.	3	.	.	.	4	.	.	.
open-hat (5)	●			●		●										
clap (10)					●								●			
kick (1)	●			●		●										

	5	.	.	.	6	.	.	.	7	.	.	.	8	.	.	.
open-hat (5)							●				●			●		
clap (10)					●								●			
kick (1)	●								●	●			●			

DANCE/EDM/HIP-HOP (Circa 1982)   130 bpm

	1	.	.	.	2	.	.	.	3	.	.	.	4	.	.	.
hat (4)	●		●	●	●		●	●	●		●	●	●	●	●	●
clap (10)					●								●			
kick (1)	●						●		●					●		

HIP-HOP (Mid-1990s)   85 bpm

	1	.	.	.	2	.	.	.	3	.	.	.	4	.	.	.
hat (4)	●		●		●		●		●		●		●		●	
clap (10)					●								●			
kick (1)	●			●		●	●					●			●	●

2.16 Drum sequencers

A drum sequencer is a tool for creating drum patterns. Early sequencers like the Roland 808 (Figure 2.7) were physical pieces of hardware. Now most people use software-based sequencers, although the basic principles are the same: Sequencers look like a grid with rows for different drum sounds and columns for short time slices (usually 16th notes or 32nd notes).

TunePad includes a drum sequencer (shown in Figure 2.15) that can be helpful for playing around with different rhythmic ideas (https://tunepad.com/composer). You can add drum sounds at different time slices by clicking on the gray squares of the grid, and once you have a pattern you like you can convert it into Python code. When converting a drum sequencer pattern to code, it can be helpful to code column by column instead of row by row. What that means is that we work left to right across the drum pattern. For each column, we look at all the sounds that hit at that time slice. We can then cue up each of those sounds using a single **playNote** statement. Here's a quick example pattern:

	1	.	.	.	2	.	.	.	3	.	.	.	4	.	.	.
clap (10)							●		●							
hat (4)			●			●				●					●	
snare (2)				●									●			
kick (0)	●		●							●	●					

We look at the first column and see that there's a single kick drum.

```
playNote(0, beats = 0.25)
```

We look at the second column and see that it's empty, so we rest:

```
rest(0.25)
```

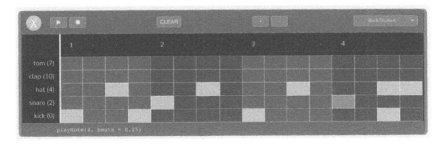

Figure 2.15 TunePad composer interface provides drum and bass sequencers.

The third column includes both a hat (note 4) and a kick (note 0). To play these together, we can use a single **playNote** command with both sounds enclosed in square brackets like this:

```
playNote([ 0, 4 ], beats = 0.25)
```

This is using a special Python structure called a list that we'll cover in more detail over the next few chapters. For now, all you need to know is that it's a convenient way to play more than one sound at the same time. If we keep going with this column-by-column strategy, here's the complete code:

```
1   playNote(0, beats = 0.25)
2   rest(0.25)
3   playNote([ 0, 4 ], beats = 0.25)  # kick + hat
4   rest(0.25)
5   playNote(2, beats = 0.25)
6   rest(0.25)
7   playNote(4, beats = 0.25)
8   playNote(10, beats = 0.25)
9   rest(0.25)
10  playNote(10, beats = 0.25)
11  playNote([ 0, 4 ], beats = 0.25)  # kick + hat
12  playNote(0, beats = 0.25)
13  playNote(2, beats = 0.25)
14  rest(0.25)
15  playNote(4, beats = 0.25)
16  rest(0.25)
```

Coding column by column can be a little quicker and produces more compact code.

Note

1 The term beat can also refer to the main groove in a dance track ("drop the beat") or the instrumental music that accompanies vocals in a hip-hop track ("she produced a beat for a new artist") in addition to other meanings.

Interlude 2
CUSTOM TRAP BEAT

In this interlude, you'll run with the skills you picked up in the preceding chapter to create a custom Trap beat. This beat will use kick drum, snare, and hi-hats. You can follow along online by visiting https://tunepad.com/interlude/trap-beat.

STEP 1: Defining variables

Start by logging into TunePad and creating a new project called "Custom Trap Beat". Add a new *Drums* instrument to your project. In this cell, declare **variables** for your drum sounds.

```
1   # variables for drums
2   kick = 1
3   snare = 2
4   hat = 4
```

STEP 2: Basic drum pattern

Here's code for a basic drum pattern. Add this code to your Drum cell after the variables:

DOI: 10.4324/9781003033240-4

```
5    # kick and snare
6    playNote(kick, beats = 0.75)
7    playNote(kick, beats = 0.25)
8    playNote(snare, beats = 1.5)
9    playNote(kick, beats = 0.75)
10   playNote(kick, beats = 0.75)
11   playNote(kick, beats = 2)
12   playNote(snare, beats = 1)
```

Let's break down each of these lines one by one:

Line	Drum name	Drum number	Beats	Python code
1	kick	1	3/4	playNote(kick, beats = 0.75)
2	kick	1	1/4	playNote(kick, beats = 0.25)
3	snare	2	1.5	playNote(snare, beats = 1.5)
4	kick	1	3/4	playNote(kick, beats = 0.75)
5	kick	1	3/4	playNote(kick, beats = 0.75)
6	kick	1	2	playNote(kick, beats = 2)
7	snare	2	1	playNote(snare, beats = 1)

When you're done, your pattern should look something like Figure 2.16.

STEP 3: Add hi-hat rolls and stutters

Now add a new **Drum Cell** to your project for the hi-hat rolls and stutters. To add our hi-hat runs let's first review **for loops** in Python (Figure 2.17).

The indented block of code is run the total number of times specified by the **range** of the loop. Try this example pattern in your project:

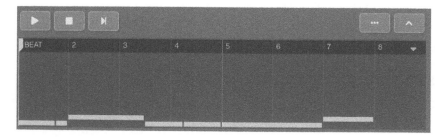

Figure 2.16 Basic drum pattern.

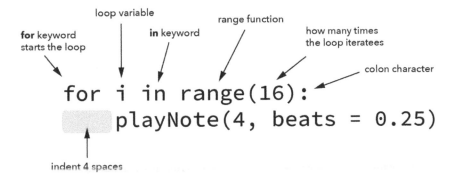

Figure 2.17 Declaring a for loop for hi-hat runs in Python.

```
1   for i in range(4):
2       playNote(hat, beats = 0.25)
3
4   for i in range(4):
5       playNote(hat, beats = 0.25 / 2)
6
7   for i in range(8):
8       playNote(hat, beats = 0.25)
9
10  for i in range(5):
11      playNote(hat, beats = 0.25 / 5)
12
13  playNote(hat, beats = 0.25)
```

Your cell should now have a pattern like Figure 2.18.

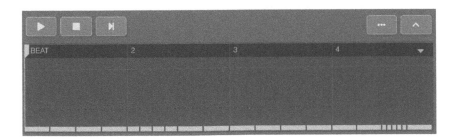

Figure 2.18 Hi-hat stutter patterns.

STEP 4: Customize

After trying the example in step 3, make up your own stutter pattern to go with your kick and snare drums. You can use any combination of beats, but make sure it adds up to a multiple of four beats so that your beat loops correctly! Here are a few for loops that play stutters at different speeds:

```
# couplet
for i in range(2):
    playNote(hat, beats = 0.25 / 2) # divide in half

# triplet
for i in range(3):
    playNote(hat, beats = 0.25 / 3) # divide into 3 parts

# quad
for i in range(4):
    playNote(hat, beats = 0.25 / 4) # divide into 4 parts

# fifthlet?
for i in range(5):
    playNote(hat, beats = 0.25 / 5) # divide into 5 parts
```

Try out different instrument sounds by changing the values of the variables and switching to a different drum kit. You can also experiment with changing the tempo. For more inspiration, this TunePad project has several popular hip-hop beat patterns that you can experiment with:
    https://tunepad.com/interlude/drum-examples.

3 Pitch, harmony, and dissonance

Chapter 2 introduced the basics of rhythm and how to use the Python programming language to code beats with percussion sounds. In this chapter, we'll explore topics of pitch, harmony, and dissonance—or what happens when you bring tonal instruments and the human voice into music. We'll start with the physical properties of sound (including frequency, amplitude, and wavelength) and why different musical notes sound harmonious or dissonant when played together. We'll also talk about different ways to represent pitch, including frequency value, musical note names, and MIDI (Musical Instrument Digital Interface) note numbers that we can use with Python code and TunePad.

3.1 Sound Waves

All sound, no matter how simple or complex, is made up of waves of energy that travel through air, water, or some other physical medium. If we could see a sound wave, it might look something like ripples of water from a pebble dropped in a still pond. The pebble is like the source of the sound, and the ripples are the sound waves that expand outward in all directions. Any source of sound (car horns, cell phone rings, chirping birds, or a plucked guitar string) sends vibrating waves of air pressure out at around 343 meters per second (the speed of sound) from the source. It's not that the air molecules themselves travel from the source of the sound to our ears; it's that small localized movements in molecules create fluctuations in air pressure that propagate outward over long distances.

DOI: 10.4324/9781003033240-5

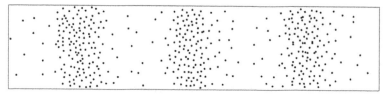

Compression wave of air molecules

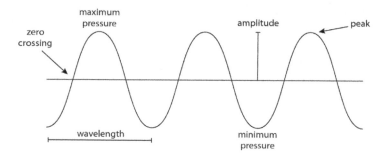

Figure 3.1 Sound is made up of compression waves of air molecules that expand outward at a speed of around 343 meters per second. The frequency of a sound wave refers to how fast it vibrates; amplitude refers to the intensity of the sound; and wavelength refers to the length of one complete cycle of the waveform.

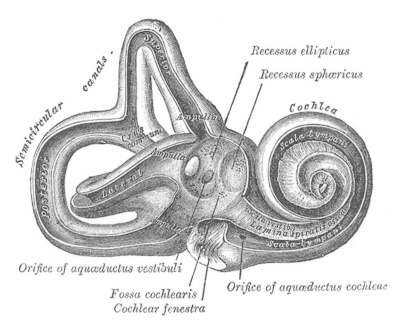

Figure 3.2 Drawing of the cochlea (inner ear).

Once those waves reach the human ear, they are captured by the outer ear and funneled to a seashell-shaped muscle in the inner ear called the **cochlea** (seen below). This muscle has tiny hairs that resonate at different frequencies causing messages to get sent to the brain that we interpret as sound.

PROTECT YOUR HEARING

As musicians or music producers, your sense of hearing is one of your most precious assets. Always wear ear protection when you're exposed to loud sustained sounds! Loud sounds can damage your inner ear permanently, meaning you can start to lose your ability to hear.

All sound waves have the following properties: **frequency, wavelength,** and **amplitude**.

3.2 Frequency

Frequency refers to the number of times a complete waveform passes through a single point over a period of time or how fast the wave is vibrating. It is measured by cycles per second in a unit called **hertz (Hz)**. One cycle per second is equivalent to 1 Hz, and 1,000 cycles are equivalent to 1,000 Hz, or 1 kHz (pronounced **kilohertz**). The higher the frequency, the higher the pitch of the sound. Figure 3.1 shows a pure sine wave, which sounds similar to a flute. The peaks and zero crossing points can both be used to mark the beginning or end of a complete cycle (Figure 3.1).

3.3 Wavelength

Wavelength refers to the length of one complete cycle of a wave in physical space. This is the distance from one peak or zero crossing to the next. We can't actually see sound waves, but the wavelength can be calculated by dividing the speed of sound (approximately 343 meters per second) by its frequency. So, for the pitch of a **Concert A** note (440 Hz), the length of the waveform would be about 78 centimeters or 2.6 feet.

$$\frac{343 \text{ m/s}}{440 \text{ Hz}} = 0.78 \text{ meters} = 78 \text{ centimeters} = 2.56 \text{ feet}$$

On most pianos, the wavelength of the lowest bass note is almost 40 feet long! In contrast, the wavelength of the highest note is only around 3 inches. The longer the wavelength, the lower the note.

Lower frequency sound also tends to travel longer distances. Think of a car playing loud music. As it approaches, you can hear the fat sound of a bass guitar or a kick drum long before you can hear other instruments. Using this property, people in West Africa were able to transmit detailed messages over long distances using a language of deep drum sounds. A drummer called a "carrier" would drum out a rhythmic pattern on a huge log drum that carried messages like "all people should gather at the market place tomorrow morning". All those within hearing range, which under ideal conditions could be as far as 7 miles, would receive the message.

3.4 Amplitude

Amplitude is related to the volume of a sound, or how high the peaks of the waveform are (Figure 3.1). You can think of this as how much energy passes through a fixed amount of space over a fixed amount of time. The human ear perceives a vast range of sound levels, from sounds that are softer than a whisper to sounds that are louder than a pain-inducing jackhammer. In order to communicate the volume of sound in a manageable way, music producers and engineers use a unit of loudness called **decibels (dB)**. The whispered voice level might be 30 dB, while the jackhammer sound would be about 110 dB. Loud noises above 120 dB can cause immediate harm to your ears.

3.5 Dynamics

The variation of amplitude levels from low to high within a musical composition is referred to as dynamics. The difference between the softest sound to the loudest sound is called the **dynamic range** of music. You can look at the **waveform** of an audio signal to get a quick sense for its dynamic range. In general, lower heights mean lower amplitude and higher heights mean higher amplitude. The loudness of a sound is also dependent on frequency. So, looking at a waveform alone won't tell you how loud something will sound to listeners (Figure 3.3).

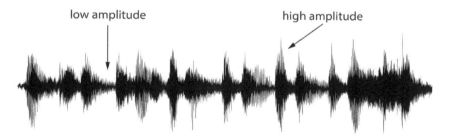

Figure 3.3 A waveform with varying amplitude.

3.6 Bandwidth

Bandwidth refers to the range of frequencies present in audio. As in the case of dynamic range, you can think of this as the difference between the highest and lowest frequencies. Humans with good hearing can distinguish sounds between 20 Hz and 20,000 Hz. Most audio formats designed for music support frequencies up to 22 kHz (pronounced 22 **kilohertz** or 22,000 hertz) so that they can capture the full range of human hearing.

Musical instruments naturally fall within the range of human hearing at different places on the frequency spectrum; this is referred to as the instrument's bandwidth. **Instrument bandwidth** is important to music producers as they arrange a musical composition. In addition to the quality of the sound of the instruments, those in different bandwidths can complement each other. Like a cello and a flute, or a bass and a saxophone. Music producers are keenly aware of the influence of low- and high-frequency instruments on their listeners. Musical instruments in the bass register are often the foundation of composition, holding everything together.

3.7 Pitch

Within the spectrum of human hearing, specific frequencies, ranges of frequencies, and combinations of frequencies are essential for creating music. This section covers some combinations of musical tones common in Western music culture. We'll then work in TunePad to try out different combinations and explore those relationships through well-known musical compositions.

While music producers and engineers often think in terms of frequencies (hertz), musicians use pitch and intervals to describe musical tones and the relationships between them. Pitches are individual notes like F, G, A, B, C, D, E as seen on the piano keyboard. The interval between each adjacent note on a traditional keyboard is called a half step or a semitone. These base pitches can also have **accidentals**. Accidentals are like modifiers to notes that raise or lower the base pitch. A note with a sharp ♯ applied has its pitch raised by a semitone, while a note with a flat ♭ applied is lowered by a semitone. The black notes on a piano are notes with accidentals. For example, moving a C♯ (black key) is a half step. Moving directly from a C to a D (both white keys) is called a whole step. Moving from a B to a C or an E to an F is also a half step because there's no black key in between (Figure 3.4).

How you refer to note names depends a little on the kind of instrument you play and its frequency bandwidth. Terms like low C, middle C, or high C might change if you're playing a piano versus a flute or if you sing soprano versus tenor. Note names repeat in groups of 12, so it's common to include the octave number along with a note's name to make things less ambiguous as shown in Figure 3.4.

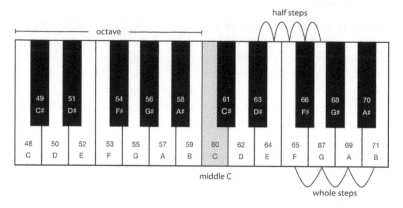

Figure 3.4 A half step is the distance between two adjacent piano keys, measured in semitones.

3.8 Musical Instrument Digital Interface

One takeaway from the previous section is that note names are confusing. There are multiple names for the same pitch (G♯ is the same as A♭), and note names are repeated every octave. To help make things less ambiguous, computers and digital musical instruments use a standardized format called **MIDI**, which stands for Musical Instrument Digital Interface. MIDI is a protocol, or set of rules, for how digital musical instruments communicate. Digital musical instruments send messages to your computer or to other musical instruments. Typical MIDI controllers look like piano keyboards or drum pads but can take many other forms as well. When you play a MIDI instrument, it sends information about a note's pitch, timing, and volume along with other messages about vibrato, pitch bend, pressure, panning, and clock signals. This table shows two octaves of notes with their typical frequency values:

Note Name	MIDI	Frequency (Hz)	Note Name	MIDI	Frequency (Hz)
C1	24	32.70	C2	36	65.40
C♯1 / D♭1	25	34.65	C♯2 / D♭2	37	69.30
D1	26	36.71	D2	38	73.42
D♯1 / E♭1	27	38.89	D♯2 / E♭2	39	77.78
E1	28	41.20	E2	40	82.41
F1	29	43.65	F2	41	87.31
F♯1 / G♭1	30	46.25	F♯2 / G♭2	42	92.50
G1	31	48.99	G2	43	97.99
G♯1 / A♭1	32	51.91	G♯2 / A♭2	44	103.83
A1	33	55.00	A2	45	110.0
A♯1 / B♭1	34	58.27	A♯2 / B♭2	46	116.54
B1	35	61.74	B2	47	123.47

The appendix contains a complete table with note names, frequency values, and MIDI numbers.

TunePad uses MIDI numbers to designate pitch. To play a C0, the lowest pitch on the TunePad keyboard, we use the code **playNote(12).** To play a C4, a middle C in the center of an 88-key piano, we use the code **playNote(60)**. MIDI notes go all the way up to note G9 with the note value 127.

Now let's experiment with pitch in TunePad. Try creating a new piano instrument in TunePad and adding this code:

```
# code for first piano cell
playNote(48)
playNote(55)
playNote(60)
playNote(55)
```

This program plays four notes: 48 is a C, 55 is a G, and 60 is a middle C. Now add a second piano instrument to the same project so that you have two cells. Add this code to the second cell:

```
# code for second piano cell
playNote(72, beats = 4) # C5
playNote(79, beats = 4) # G5
playNote(76, beats = 4) # E5
playNote(79, beats = 4) # G5
```

This Python program looks similar to the first one, but we've changed the length of each note using the **beats** parameter. In this case, we're asking TunePad to play four notes, each four beats long. Try playing both piano parts at the same time. We can also make our notes shorter instead of longer. Let's add a third piano instrument with notes that are each one half beat long. Try playing all three pianos together.

```
# code for third piano cell
playNote(36, beats = 0.5)
playNote(36, beats = 0.5)
playNote(43, beats = 0.5)
playNote(43, beats = 0.5)
playNote(48, beats = 0.5)
playNote(48, beats = 0.5)
playNote(43, beats = 0.5)
playNote(43, beats = 0.5)
```

3.9 Harmony

Harmony in music can be defined as a combination of notes that, when played together, have a pleasing sound. Although opinions about what sounds good in music are highly subjective, certain combinations of notes played together can elicit predictable psychological responses—some combinations of notes sound **harmonious** while others some **discordant**. Musicians use this phenomenon to create an emotional tone for their compositions.

In Western music, much of our conception of pitch is built on different mathematical ratios. Consider the string of an instrument like a guitar or violin. Plucking the open A (second lowest) string plays an A, which has a frequency of 110 Hz. Now if we touch the string at its midpoint, dividing it in half, we still hear an A an octave above the previous one—twice the frequency of the first note, or 220 Hz. If we touch the string a third of the way down and pluck it, the result is an E above the higher A. This E is exactly three times our original frequency, or 330 Hz. Likewise, dividing the string into fourths multiplies the original frequency by four. We can continue this division on the string as follows (Figure 3.5):

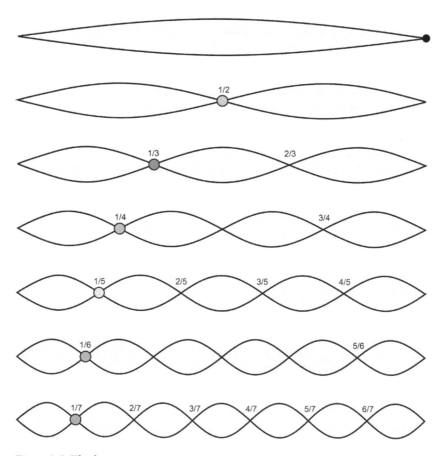

Figure 3.5 The harmonic series.

The resulting sequence of ascending pitches this produces is known as the **harmonic series.** If two notes have a harmonic relationship, that means that the two frequencies have an integer relationship. Basically, that means there is a way to divide the two frequencies such that the result is a whole number. The harmonic series is simply the set of frequencies that have a harmonic relationship to a **fundamental pitch** (the initial note). Our initial experiment with the string illustrates this relationship.

To find the frequencies that make up the harmonic series for a given pitch, we multiply its frequency by the set of whole numbers. For A1, which is 55 Hz, the first eight harmonics would be the following:

Harmonic	Frequency	Note name	MIDI
1	55 Hz	A1	33.00
2	110 Hz	A2	45.00
3	165 Hz	E3	52.00
4	220 Hz	A3	57.00
5	275 Hz	C#4	60.86
6	330 Hz	E4	64.02
7	385 Hz	G4	66.69
8	440 Hz	A4	69.00

In the table above, the MIDI value is given for each harmonic of A1. Notice that these values are given in decimal format. In TunePad, `playNote` accepts both whole and decimal values. The whole numbers are a data type referred to as **integer** values, or just as **ints**. The decimals are a separate data type referred to as **floating point** values, or just **floats**.

Listen to an example here: https://tunepad.com/examples/harmonic-series.

Notes with frequencies that form simple ratios, such as 2:1, 3:2, 4:3, or 5:4, tend to sound good together. For instance, we can take the note A4 (440 Hz) and add a frequency that is 1.5 times its value, giving us an E4 (660 Hz). This results in a ratio of 3:2 and a pleasant sound. However, if we add a frequency that is 1.3 times the value of 440 Hz, we end up with 572 Hz, which creates a not-so-pleasant combination of tones. It's not an accident that there is no corresponding musical note to 572 Hz on the piano keyboard.

3.10 Intervals

In music, an **interval** is the distance between two notes. These notes can either be played simultaneously or not. If they are played simultaneously, the pitches are called a **dyad** or a **chord**. Otherwise, they are a **melodic** interval. An interval is always measured from the lowest note. Intervals have two different components: the **generic interval** and the quality. The generic interval is the distance from one note of a scale to another; this can also be described as the number of letter names between two notes, including both notes in question. For example, the generic interval of C4 and E4 has C4, D4, and E4 in between. That's three notes, so we have a third. The generic interval between F#3 and G3 is a second. The generic interval between G2 and G3 is an eighth—also known as an octave. The quality can be one of five options: Perfect, Major, Minor, Augmented, or Diminished. Each quality has a distinct sound and can generate different emotional responses. Here are some common intervals in music along with their frequency ratios and half steps.

Ratio	Interval name		Half steps
2:1	Octave		12
15:8	Major Seventh		11
16:9	Minor Seventh		10
5:4	Major Sixth		9
8:5	Minor Sixth		8
3:2	Perfect Fifth	😇	7
45:32	Tritone	😈	6
4:3	Perfect Fourth		5
5:4	Major Third		4
6:5	Minor Third		3
9:8	Major Second		2
16:15	Minor Second		1
1:1	Unison		0

These intervals are based on the harmonic series, but this isn't exactly how most instruments are tuned. We'll talk more about this below. Also, the naming of the ratios (Fifth, Fourth, Major Third, etc.) will make more sense in Chapter 5 where we talk about scales and keys. Notice that there's one particularly nasty-looking ratio called the Tritone interval (45:32). This interval has historically been referred to as the *Devil in Music* and was frequently avoided in music composition for its dissonant qualities.

One of the simplest and most common intervals is the octave, which has a frequency ratio of two to one (2:1)—meaning that the higher pitch completes two cycles in the same amount of time that the lower pitch completes one full cycle. Notes that are octave intervals from one another have the same letter name and are grouped together on a piano keyboard. Notice the repeating patterns where C is the highlighted note, C3, C4, C5 (Figure 3.4). To illustrate, we can begin with **middle C** (C4), which is approximately 262 Hz, and then move to a C5, which is an octave above it at 524 Hz. You can see that C5 is double the C4 frequency forming the octave ratio, 2:1. The waveforms representing the two notes forming this octave are plotted in Figure 3.6. We can see that for every single complete cycle of the 262 Hz wave, C4, there are two full cycles of the waveform for the octave above it, 524 Hz C5. It's easier to count cycles if you look at the zero-crossings.

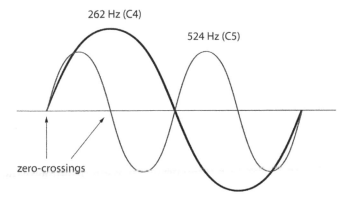

Figure 3.6 Two waves at an interval of an octave.

What does an octave look like in code? As we have seen, each note on the piano keyboard is a half step, and there are 12 half steps between octaves. Try counting the notes between C4 and C5. Remember, the black keys count!

TunePad tracks the notes on the keyboard by half steps, so you can easily play any octave interval without having to figure out the exact note number. For instance, this code plays a middle C (60) and a C one octave higher.

```
note = 60
playNote(note)
playNote(note + 12)
```

This code assigned the number 60 to the variable **note** on the first line. The third line played a note one octave higher by adding 12 to the original **note** variable (note 72 is played). Expanding on this, you can substitute any number you want for **note** and generate an octave above it by adding 12.

Octaves sound good together in music and are used in many popular songs. For instance, in the song *Over the Rainbow* composed by Harold Arlen from the *Wizard of Oz*, the beginning two notes are an octave apart.

```
1    # First two bars of "Over the Rainbow"
2    # Composed by Henry Arlen
3    playNote(60, beats = 2) # note C4
4    playNote(60 + 12, beats = 2) # note C5
5    playNote(71, beats = 1)
6    playNote(67, beats = 0.5)
7    playNote(69, beats = 0.5)
8    playNote(71, beats = 0.5)
9    rest(0.5)
10   playNote(72, beats = 1)
```

Try this example at https://tunepad.com/examples/rainbow.

Another interval relationship important to Western music is the ratio of 3:2, also known as the perfect fifth, which has seven half steps between notes. With this interval ratio, there are three complete cycles of the higher frequency for every two periods of the lower frequency (Figure 3.7).

Henry Mancini uses a perfect fifth (G3 392 Hz and D4 587 Hz) in the first two notes in the melody for the song *Moon River*.

We'll code the first few bars of *Moon River* using a variable called **root_note** to set the starting note. This allows us the flexibility to easily play the song beginning from any note on the keyboard and the relationship between notes stays the same no matter which note you start with. Try changing the value of the variable **root_note** to another MIDI note. This can come in handy when you are composing for a singer who would rather have the song in another key or octave.

```
1    # First bars of "Moon River"
2    # Composed by Henry Mancini
3    root_note = 55
4    playNote(root_note, beats = 3)
5    playNote(root_note + 7, beats = 1)
6    playNote(root_note + 5, beats = 2)
7    playNote(root_note + 4, beats = 1.5)
8    playNote(root_note + 2, beats = 0.5)
9    playNote(root_note, beats = 0.5)
10   playNote(root_note - 2, beats = 0.5)
11   playNote(root_note, beats = 2)
```

See this example at https://tunepad.com/examples/moon.

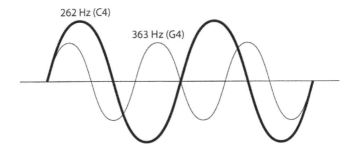

Figure 3.7 The ratio between the note C 262 Hz and the note G 393 Hz is considered a perfect fifth.

3.11 Dissonance

Dissonance refers to combinations of notes which when combined have an unpleasant sound that creates tension. The interval of a minor second (or one half step) is a complex frequency ratio of about 9.5:1. This combination gives you a sense of suspense. You can hear the effect of dissonance used in the composition by John Willimas for the movie *Jaws*. We'll use a for loop for this example, as the two notes are repeated.

```
1   # bass line for the theme from Jaws
2   # composed by John Williams
3   for i in range(8):
4       playNote(40, beats = 0.5) # E2
5       playNote(41, beats = 0.5) # F2
```

Intervals that are dissonant are unstable, leaving the listener with the impression that the notes *want* to move elsewhere to resolve to more stable or **consonant** intervals.

You can try this example in TunePad to hear how the notes that are one half step apart crunch when played together.

```
1    # half step - the notes are just 1 number away
2    playNote(41, beats=1, sustain=3)
3    playNote(42, beats=1, sustain=2)
4    rest(2)
5    playNote([41, 42], beats=4)
6    rest(2)
7
8    # whole step - these notes are 2 numbers away
9    playNote(41, beats=1, sustain=3)
10   playNote(43, beats=1, sustain=2)
11   rest(2)
12   playNote([41, 43], beats=4)
13   rest(2)
```

Listen to this example here https://tunepad.com/examples/dissonance.

Another example of the use of dissonant intervals comes from the horror movie *Halloween* (1978). The theme song by John Carpenter creates a sense of suspense and deep unease with the use of dissonant intervals such as the Tritone (ratio 45:32).

3.12 Temperaments and Tuning

Follow along at https://tunepad.com/examples/temperaments.

The intervals in the previous sections were based on ratios called perfect or pure intervals. The waves of so-called perfect intervals align at a simple integer ratio. If two tones form a perfect interval, it will result in a louder sound, as the amplitudes are added. If one of the tones is out of tune, then there will be interference between the two waves. This interference manifests as an audible rhythmic swelling or "wah-wah" between the waves, which we call **beating**. The farther the two tones are from being perfect, the faster the beating. If the tones are apart far enough, you might even hear this beating as a third tone, called a **combination tone**. Pure and impure intervals are not a value judgment but a description of natural phenomena.

Using notes based on these simple ratios seems to make a lot of sense—it's based on simple mathematical relationships that we know sound good to the human ear. But, it turns out that we quickly run into problems using this system when we start trying to tune an instrument like a piano. For example, say we're trying to tune an A4 against a fixed lower tone on a keyboard using pure ratios. If we're tuning this A4 against an F4 at approximately 349 Hz, our intervals form a major third at a 5:4 ratio of frequencies. This results in our A4 being approximately 436.26 Hz. But, if we tune our A4 against an F#4 at 370 Hz, this produces a minor third, which is at a 6:5 ratio of frequencies. Now our A4 is 444 Hz instead of 436.25 Hz! How can it be that the same note maps to different frequencies?

The question of how to map frequency—of which there are endless possible values—to a finite set of notes means that we have to both arbitrarily choose a starting point and also decide at what intervals to increment. This is the basis for what are called temperaments. Temperaments are systems that define the sizes of different intervals—how the tones relate to one another. In choosing the tones in an octave, we must compromise between our melodic intervals and our harmony. Ideally, we want a system with as consistent melodic intervals and that is as close to perfect harmonic intervals as possible. In a system based on perfect ratios—also referred to as **Just Intonation**—the divisions, or semitones, of an octave are not evenly distributed. This means that there are unique sets of tunings for every note we choose as the base note of our octave. Just Intonation also does not form a closed loop of an octave. This is getting into the weeds a bit, but if we derive each note's frequency by tuning the ratio of a perfect fifth (3:2) from the previous note, we do not end up at the same place one octave higher. In fact, tuning the ratio of 3:2 twelve times brings us back to our exact starting note only after seven octaves. Because Just Intonation has too many mathematical snares to be represented by the 12 notes of the keyboard, it's not a stable tuning system and not a temperament. By definition, a temperament is a calculated deviation from Just Intonation that maps each note to exactly one frequency while still getting as close as possible to the pure intervals.

Most contemporary music, including TunePad, is based on a system called Equal Temperament. The octave at a pure 2:1 ratio serves as the foundation, which is then divided into 12 equal half steps. Most often, Western harmony is built primarily from thirds, fifths, and octaves. Every octave (and unison) is a pure interval in Equal Temperament. The perfect fourths and fifths are *nearly* pure intervals. Major and minor thirds are quite far from perfect, but because we have grown so accustomed to hearing these intervals, they do not sound off to our ears. Because Equal Temperament is, well, equal, every chord will have the same sound in every key. Each semitone is equally sized, and every note maps to exactly one frequency. Furthermore, each semitone is divided into 100 cents, which we can use to further specify intonation. With our intervals decided, we now only have to choose a starting pitch from which to tune the others. Most of the time in North America, the system is aligned to A440, meaning that A4 is equal to exactly 440 Hz.

Keyboard instruments have fixed pitch, while singers and instruments such as violin or flute have flexible tuning. In acoustic performance, pitch can vary due to many factors. No instrument is perfectly in tune. Tuning can be affected by factors such as air pressure and temperature. Even a performer's physiology can affect tuning. Often, performers will tune harmonies using Just Intonation such that a chord uses pure intervals and is more pleasing. Many musicians will do this without even being aware that they are doing it—Just Intonation just *feels* in tune.

It's important to remember that the decision to tune to A440 and to divide the octave into 12 equal semitones is only one possibility in response to debates about musical tuning that date back thousands of years, and it's only one of a myriad of ways that music can be tuned. There are many alternate tuning systems, both historical and contemporary from both Western and non-Western cultures, which are still in use today.

Interlude 3

MELODIES AND LISTS

For this interlude we'll code a short section of a remix of Beethoven's *Für Elise* created by artist and YouTuber Kyle Exum (Bassthoven, 2020). Because this song has a more intricate melody, we'll learn how to play sequences of notes written out as Python **lists**. We'll talk more about lists in the next chapter, but for now, you can think of them as a way to hold more than one note in a single variable.

STEP 1: Variables

Create a new Keyboard instrument and add some variables for our different note names.

```
A = 69      # set variable A equal to 69
B = 71
C = 72
D = 74
E = 76
Eb = 75     # E flat
Gs = 68     # G sharp
_ = None
```

The last line is a little strange. It defines a variable called _

- In Python, the underscore _ character is a valid variable name.
- We set this variable to have a special value called **None**.
- Calling **playNote** with this value is the same thing as a rest. It plays nothing.

DOI: 10.4324/9781003033240-6

STEP 2: Phrases

- For this song, we're going to define four musical phrases that get repeated to make the melody.
- Each phrase gets its own variable.
- Each variable will hold lists of notes in the order they should be played.
- You create a Python list by enclosing variables inside of square brackets (see the code below).
- We use the underscore character _ mean play nothing.
- Sometimes we subtract 12 from a note. That means to play the note an octave lower.

```
# four basic phrases that repeat throughout

p1 = [ E, Eb, E, Eb, E, B, D, C, A, _, _, _ ]
p2 = [ A, C - 12, E - 12, A, B, _, _, _ ]
p3 = [ B, E - 12, Gs, B, C, _, _, _, C, _, _, _ ]
p4 = [ B, E - 12, C, B, A, _, _, _, A, _, _, _ ]
```

STEP 3: Playing the Phrases

Now that we've defined our variables, we can start to play the melody. One way to do this is to use a Python **for loop** to iterate through every note. One cool thing about Python is that we can join lists together using the plus sign (+). Here's what everything looks like together. Don't worry about understanding all of the details yet. We'll go through all of these concepts in the coming chapters.

```
1   _ = None
2   A = 69
3   B = 71
4   C = 72
5   D = 74
6   E = 76
7   Eb = E - 1      # E flat
8   Gs = A - 1      # G sharp
9
10  # four basic phrases that repeat throughout
11  p1 = [ E, Eb, E, Eb, E, B, D, C, A, _, _, _ ]
12  p2 = [ A, C - 12, E - 12, A, B, _, _, _ ]
13  p3 = [ B, E - 12, Gs, B, C, _, _, _, C, _, _, _ ]
14  p4 = [ B, E - 12, C, B, A, _, _, _, A, _, _, _ ]
15  p5 = [ A,_,_,_,A,_,_,_,A,_, _, _, A, _, _, _ ]
16
17  for note in p1 + p5 + p2 + p3 + p1 + p2 + p4:
18      playNote(note, beats = 0.5)
19
20  for note in p1 + p2 + p3 + p1 + p2 + p4:
21      playNote(note, beats = 0.5)
```

STEP 4: Bass!

Add a Bass to your project and change the voice to **808 Bass**. You can copy the code below for the bass pattern (Figure 3.8):

STEP 5: Drums

To finish up, let's layer in a simple drum pattern that works well with the melody. Create a new **Drum instrument** and add this code.

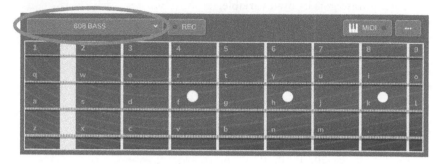

Figure 3.8 Select the 808 Bass voice.

```
1    rest(12)
2    for i in range(4):
3        playNote(21)
4        rest(2)
5        playNote(21)
6        rest(1)
7        playNote(16, beats = 0.5)
8        rest(1)
9        playNote(16, beats = 0.5)
10       rest(1)
11       playNote(21)
12       rest(2)
13       playNote(21)
14       rest(1)
15       playNote(28, beats = 0.5)
16       rest(1)
17       playNote(16, beats = 0.5)
18       rest(1)
```

```
1    for i in range(16):
2        playNote(0)
3        playNote(2, beats = 0.5)
4        playNote(2, beats = 0.5)
5        playNote(10)
6        playNote(0)
7
```

You can try this project online: https://tunepad.com/interlude/bassthoven.

4 Chords

Chords are an essential building block of musical compositions. The skillful use of chords can set the foundation of a song and create a sense of emotional movement. However, even though the basic ideas behind chords are easy to understand, there's an overwhelming amount of terminology and technical detail that can take years to learn. Using code helps us cut through layers of complicated terminology to reveal elegant structures beneath. With code, chords are nothing more than lists of numbers that follow consistent patterns. We'll work through chords using Python **lists** and **functions**. You'll learn some of the traditional music terminology and what it means, but you'll also build your own toolkit of computer code to use for new compositions.

4.1 Chords

You can follow along with interactive online examples at https://tunepad. com/examples/chord-basics

In Chapter 3, we introduced the idea of harmony and dissonance. Two or more notes have a harmonic relationship if their frequencies have integer ratios. For example, when two notes are one octave apart, the higher note vibrates exactly two complete cycles for every one cycle of the lower note (a 2:1 ratio). When two notes are a fifth apart, their frequencies have a 3:2 ratio. The higher note vibrates exactly three times for every two complete cycles of the lower note.

DOI: 10.4324/9781003033240-7

Building on this idea of harmonic relationships between notes, a **chord** is more than one note played together at the same time. In Python we can think of a chord as a **list** of numbers representing MIDI (Musical Instrument Digital Interface) note values. For example, this code plays a C major chord in TunePad.

```
Cmaj = [ 48, 52, 55 ] # notes C, E, G
playNote(Cmaj)
```

Here **Cmaj** is a variable. Instead of assigning that variable to a single number, we're assigning it a *list* of numbers. In Python, a list is a set of values enclosed in square brackets and separated by commas. Then on the second line we play all three notes together using the **Cmaj** variable (Figure 4.1).

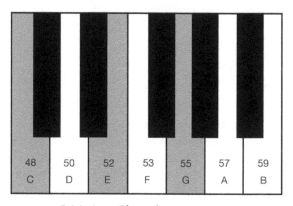

C Major Chord

Figure 4.1 C major chord with MIDI note numbers.

We can also play the same chord using just a single line of code where we pass the list of numbers directly to the **playNote** function.

```
playNote([ 48, 52, 55 ])
```

But, using the variable is nice because it helps make our code easier to read and understand. Here are a few other chord examples:

```
Cmaj = [ 48, 52, 55 ] # C major chord
Fmaj = [ 53, 57, 60 ] # F major chord
Gmaj = [ 55, 59, 62 ] # G major chord
```

A chord's name comes from two parts. The first part is the **root** note of the chord—usually the first note in the list. For example, an F major chord starts with note 53 (or an F on the piano keyboard). And, the G major chord starts with the note 55, a G on the keyboard.

The second part of a chord's name is its type or **quality**. In our examples, **Cmaj**, **Fmaj**, and **Gmaj** are all *major* chords. Later in this chapter we'll review several common chord types and how to create them in code. Each chord type has a consistent pattern. For example, all major chords follow the exact same pattern: take the root note, add 4 to get the second note, and then add 7 to get the last note. You can build a major chord up from any base note you want as long as it follows this same pattern (Figure 4.2).

Figure 4.2 Creating chords as lists of numbers in Python. Each major chord follows the same pattern.

Another way to write this in code is to define a single root note variable and then create the chord based on the root:

```
root = 48
playNote([ root, root + 4, root + 7 ]) # C major
root = 52
playNote([ root, root + 4, root + 7 ]) # F major
```

One thing to notice about this code is that the first two lines and the last two lines are almost identical. We're just changing the root note value. In fact, all we need to make any chord we want is its root note and the pattern that defines its quality. Once we know the patterns, the rest is easy.

But, it would be tedious and error prone to write out [root, root + 4, root + 7] every time we wanted to use a major chord. Fortunately, Python gives us a powerful tool for exactly this kind of situation: **user-defined functions**.

4.2 User-defined functions

In the first few chapters of this book we've been using functions to play music: **playNote**, **rest**, and **moveTo** are all functions provided by Tune-Pad. When we want to use a function, we just type its name and list the parameters to send it inside parentheses. With Python, we can also create our own functions to build up a musical toolkit. Creating functions also helps make code shorter and easier to understand because we'll be able to use the same segments of code over and over again without having to copy and paste. Here's a quick example that creates a major chord based on a root note.

```
1  def majorChord(root):
2      chord = [ root, root + 4, root + 7 ]
3      return chord
```

And, now that we've defined the function, we can use it as a shortcut in TunePad.

```
4  Cmaj = majorChord(48)
5  Fmaj = majorChord(53)
6  playNote(Cmaj, beats = 2)
7  playNote(Fmaj, beats = 2)
```

There's a lot going on with these few lines of code. Let's break it down line by line.

- Line 1 starts with the **def** keyword. This is short for "define", and it tells Python that we're about to define a function.
- Next is the name of the function we're defining. In this case, we're calling it **majorChord**, but we could use any name want as long as it follows Python's naming rules.[1]
- After the function name, we need to list out all of the function's **parameters** enclosed inside parentheses. In this case, there's only one parameter called **root**. If you need more than one parameter, you separate each parameter with commas. We'll talk about this more later on in the book, but you can think of parameters as special kinds of variables that are only usable inside of a function.
- After the parameter list, we need the colon character (:). This tells Python that the **body** of the function is coming next (Figure 4.3).

Figure 4.3 How to declare a user-defined function in Python.

- Line 2 starts the body of the function. Here we're just creating a variable called **chord** and assigning it to a list of numbers that define a major chord. The numbers are the root note, the root note + 4, and the root note + 7. An important thing to notice is that this line of code is indented by four spaces. Just like with loops, indentation tells Python that these lines are part of the function body. They're considered to be inside the function, not outside.

- Line 3 is also indented by four spaces because it's also part of the function body. This line uses the special **return** keyword to say what value the function produces. In this case, we're returning the **chord** variable, the list of three numbers that make up our major chord. When Python gets to the **return** keyword, it immediately exits the function and returns the value given on that line.
- It's also possible to define a function with no return value. In this case, Python provides a special return value called **None**.

That's it for our function definition. Now we can see how it's used on lines 4–7. Notice that we can call our new function multiple times to generate different chords, letting us reuse code to create more readable and elegant programs. In the rest of the chapter, we'll use this same template to begin to build up a library of chord types, each with its own function.

You can define variables inside a function just like you might otherwise. A variable's **scope** refers to where we can use this variable or function. You can think of this as the level of indentation for a function. Variables defined within the definition of a function can only be used in that function; their scope is within that function. The following code will result in an error on line 4.

```
1  def majorChord(root):
2      chord = [ root, root + 4, root + 7 ]
3  majorChord(60)
4  playNote(chord) # ERROR
```

The **chord** variable lives only in our **majorChord** function. We can refer to this as a local variable. Alternatively, there are global variables which can be used anywhere after they've been defined. In the code below, **root** and **chord** are global variables, and therefore can be freely used in the body of any functions:

```
1  root = 60
2  chord = [ root, root + 4, root + 7 ]
3  def songPar1():
4      playNote(chord)
5      playNote(root)
6  songPart1()
```

Before we move on, let's define one more function that shows how we can use our new functions inside of other functions:

```
def playMajorChord(root, duration):
    chord = majorChord(root)
    playNote(chord, beats = duration)
```

This function uses our **majorChord** function to both build a chord from a root note and then calls **playNote** to play the chord. This new **playMajorChord** function takes two parameters, the **root** note and a **duration** that says how long to play the note.

Note: By convention, Python function and variable names use lowercase letters with different words separated by the underscore character (e.g. `play_major_chord`). This style is referred to as "snake_case". However, this book uses a different style called "camelCase" where names start with a lowercase letter and uppercase letters are used to start new words (as in `playMajorChord`). We use camelCase in this book to be consistent with other programming language conventions such as JavaScript, but you should feel free to use snake_case for your own Python programs if you want to.

4.3 Common chord types

This section reviews some of the most commonly used chord types in modern musical genres. For each chord we'll provide the pattern of numbers that defines its quality, an example of a chord of that type on the piano keyboard, and a TunePad function that generates chords from a root note. We'll also describe the chord in terms of musical intervals. The names for intervals can be a little confusing, especially when combined with note names or MIDI note values.

Common chord types/qualities

Major triad	Major Seventh	Suspended 2
Minor	Minor Seventh	Suspended 4
Diminished	Dominant Seventh	Augmented

To hear these chords in action, go to https://tunepad.com/examples/chord-functions.

4.3.1 Major triad

Major chords are commonly described as cheerful and happy. They consist of a root note, the root + 4, and the root + 7. In music theory, the second note of a major triad is called a major third, and the third note is called a perfect fifth. This can be referred to as a **triad** due to the fact there are three distinct notes (Figure 4.4).

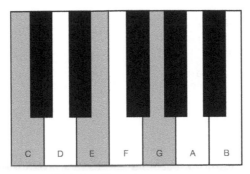

C Major Chord

Figure 4.4 C major chord.

Pattern: [0, 4, 7]
Intervals: Major 3rd, Perfect 5th
Notation: C major, CMaj, CM, C
Python Function:

```
def majorChord(root):
    return [ root, root + 4, root + 7 ]
```

4.3.2 Minor triad

Minor chords convey more of a somber tone. They're very similar to major chords except that the second note adds 3 to the root note instead of 4. In music theory, the second note is called a minor 3rd (instead of a major 3rd). But, even with this small change, the difference in mood is dramatic (Figure 4.5).

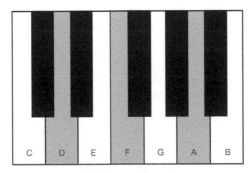

D Minor Chord

Figure 4.5 D minor chord.

Pattern: [0, 3, 7]
Intervals: Minor 3rd, Perfect 5th
Notation: D minor, Dmin, Dm
Python Function:

```
def minorChord(root):
    return [ root, root + 3, root + 7 ]
```

4.3.3 Diminished triad

Diminished chords instill tension and instability in music. They're often used as a way to transition between chords in a progression. There are a few different types of diminished chords, but the simplest, the three-note variety, is almost identical to the minor triad except that the last note is decreased (diminished) by 1. This is called a diminished 5th interval (Figure 4.6).

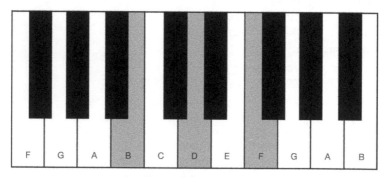

B Diminished Chord

Figure 4.6 B diminished chord.

Pattern: [0, 3, 6]
Intervals: Minor 3rd, Diminished 5th
Notation: B dim, B°
Python Function:

```
def diminishedTriad(root):
    return [ root, root + 3, root + 6 ]
```

4.3.4 Major 7th

The major 7th chord starts with a major triad and then adds a fourth note to the end of the list (root + 11). This addition is called a major 7th interval. The extra note creates a more sophisticated and contemplative feeling (Figure 4.7).

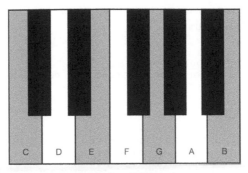

C Major 7 Chord

Figure 4.7 C major 7th chord.

Pattern: [0, 4, 7, 11]
Intervals: Major 3rd, Perfect 5th, Major 7th
Notation: Cmaj^7, CM^7, CMa^7
Python Function:

```
def major7th(root):
    return [ root, root + 4, root + 7, root + 11 ]
```

4.3.5 Minor 7th

A minor 7th starts with a minor triad and adds a minor 7th (root + 10). This chord is a bit more moody than the major 7th in feel (Figure 4.8).

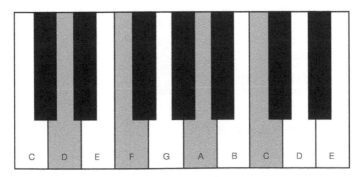

D Minor 7 Chord

Figure 4.8 D minor 7th chord.

Pattern: [0, 3, 7, 10]
Intervals: Minor 3rd, Perfect 5th, Minor 7th
Notation: Dmin^7, Dm^7
Python Function:

```
def minor7th(root):
    return [ root, root + 3, root + 7, root + 10 ]
```

4.3.6 Dominant 7th

Just like the major and minor 7ths, we can create a dominant 7th by combining some of our earlier building blocks. The dominant 7th starts with a major triad and adds a minor 7th to get the pattern [0, 4, 7, 10]. The combination of major and minor intervals can create a feeling of restlessness (Figure 4.9).

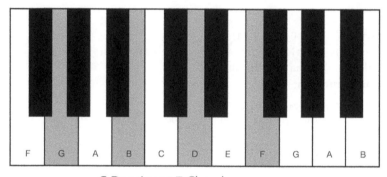

G Dominant 7 Chord

Figure 4.9 G dominant 7th chord.

Pattern: [0, 4, 7, 10]
Intervals: Minor 3rd, Perfect 5th, Minor 7th
Notation: G^7
Python Function:

```
def dominant7th(root):
    return [ root, root + 4, root + 7, root + 10 ]
```

4.3.7 Suspended 2 & suspended 4

Suspended triad chords start with a major triad, but shift the middle note up or down. A sus2 chord combines a root note, a major 2nd (+2), and a perfect fifth (Figure 4.10).

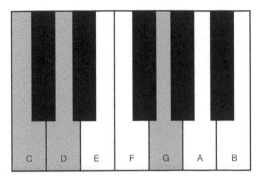

C sus 2 Chord

Figure 4.10 Csus2 chord.

Pattern: [0, 2, 7]
Intervals: Major 2nd, Perfect 5th
Notation: Csus2, C^{sus2}
Python Function:

```
def sus2(root):
    return [ root, root + 2, root + 7 ]
```

A sus4 chord shifts the middle note in the other direction to a major 4th (+5) (Figure 4.11).

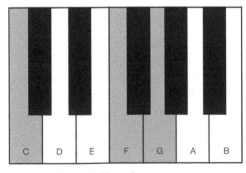

C sus 4 Chord

Figure 4.11 Csus4 chord.

Pattern: [0, 5, 7]
Intervals: Major 4th, Perfect 5th
Examples: Csus4, C^{sus4}
Python Function:

```
def sus4(root):
    return [ root, root + 5, root + 7 ]
```

You can try both versions of the suspended chord in the interactive tutorial.

4.3.8 Augmented triad

The last type of chord we'll cover here is called an augmented triad. This is just a major triad with a "sharpened 5th": the last note is raised from a perfect 5th (+7) to an augmented 5th (+8). This chord can add a feeling of suspense or anxiety (Figure 4.12).

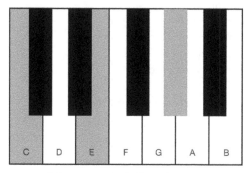

C Augmented Chord

Figure 4.12 C augmented chord.

Pattern: [0, 4, 8]
Intervals: Minor 3rd, Perfect 5th
Notation: Caug, C+
Python Function:

```
def augmentedTriad(root):
    return [ root, root + 4, root + 8 ]
```

There are many, many other kinds of chords we can explore that add extra notes or use notes in different patterns. Extended chords bring in intervals like 9ths, 11ths, and 13ths, and inverted chords shift the position of the root note. With 88 keys on a piano keyboard and dozens of chord qualities to choose from, there are hundreds and hundreds of possible chords we can use in a given song. How do we reduce this complexity to generate music that sounds good? There are three answers to this question:

- First, **musical keys** are like templates that give us a collection of chords and notes that will sound good together. Once we know what key we're in, the set of possible chords becomes much more manageable. The next chapter will cover the main ideas behind keys and scales.
- Second, there are standard **chord progressions** that are used consistently in different genres of music. A chord progression is a sequence of chords that set up a compositional structure for a piece of music. In Chapter 6, we'll show how to generate progressions of chords from common templates.
- The third answer is simply hard-won experience. As you develop your musical ear, you'll become more and more familiar with chord types and progressions and how they're used in different genres. This experience will help you begin to innovate and improvise.

Common interval names	
Semitones	*Name*
+ 0	Unison
+ 1	Minor 2nd
+ 2	Major 2nd
+ 3	Minor 3rd
+ 4	Major 3rd
+ 5	Perfect 4th
+ 6	Tritone
+ 7	Perfect 5th
+ 8	Minor 6th
+ 9	Major 6th
+ 10	Minor 7th
+ 11	Major 7th
+ 12	Octave
+ 13	Minor 9th
+ 14	Major 9th
+ 15	Minor 10th

Note

1 Function names have to start with a letter (lowercase or uppercase). Names can include letters, numbers, and the underscore _ character. Unicode characters are also allowed (so something like 플레이 노트 can also be a valid Python function name).

Interlude 4

PLAYING CHORDS

In this interlude we're going to explore a few options for playing chords using TunePad and Python. When you're composing the harmony of your song, you have more to consider than just what chords to choose. You also have to consider how to play these chords. Subtle variation in timbre, harmonics, and timing can make a huge difference in the sound that you ultimately produce. In Chapter 4, we saw how to play chords using a list and a single **playNote** statement like this:

```
playNote([48, 53, 55], beats = 4.0)
```

This is your most basic and most mechanical sounding option. Here are a few other ideas and techniques to experiment with. You can follow along online with this TunePad project: https://tunepad.com/interlude/play-chords

OPTION 1: Block chords

With block chords you play every note in a chord at the exact same time and for the exact same duration. This approach is simple and can add a strong rhythmic feel to your music. But in some situations using block chords can sound harsh and overly mechanical. Here's a simple function that takes a chord (a list of numbers) and plays each note at the same time for an equal duration:

DOI: 10.4324/9781003033240-8

```
1  def block(chord, beats):
2      playNote(chord, beats)
```

OPTION 2: Rolled chords

Sometimes when humans play chords, they introduce subtle variations in the timing between note onsets. This variation can be intentional and exaggerated or simply a natural result of playing or strumming chords by hand. This style is called a **rolled** chord. We might roll a chord if you want to bring out a change in harmony or if you want to emulate a strummed instrument. Artists like Dr. Dre have used rolled chords on the piano to create iconic sounds. You can also combine rolled chords and block chords. If your chord progression is changing chords, you can draw attention to this by rolling the chord that changes. This function rolls a chord by adding a short, fixed delay between each note onset.

```
1  def rolled(chord, duration):
2      delay = 0.1   # how far to space out note start times
3      offset = 0    # accumulated delay
4      for note in chord:
5          playNote(note, beats = delay, sustain = duration - offset)
6          offset += delay   # keep track of accumulated delay
7      fastForward(duration - offset)
```

The function uses a couple of "bookkeeping" variables called **delay** and **offset**. The **delay** variable just says how long to pause before each successive note in the chord is played. The **offset** variable keeps track of the total amount of delay that we've introduced in the for loop. We use the **offset** variable in two places. First, on line 6, we adjust the **sustain** parameter so that all of the notes in the chord are released at the exact same time (the sustain parameter lets the note ring out longer than what gets passed in the beats parameter). Second, on line 9, we adjust the playhead position after the loop finishes. This makes it so that calling the rolled function advances the playhead forward by the *exact* amount specified in the **duration** parameter. One other quick note: line 7 uses the plus–equal operator (+=). This is a shorthand way of saying:

```
offset = offset + delay.
```

OPTION 3: Random rolled chords

We can take this technique a step further by *randomly* varying the note onset times. To do this, we'll use one of Python's utilities from the random module: the **uniform** function. This function takes an input of two numbers and generates a random decimal number between those two numbers. We'll use this to generate an offset between each successive note in the chord. As with the previous example, we'll use the **sustain** parameter of **playNote** to hold out each note for the remaining duration of the original inputted beats, subtracting the total cumulative amount of offset each time.

```
1   from random import uniform
2   def rolled(chord, duration):
3       max_delay = 0.15
4       offset = 0
5       for note in chord:
6           next_delay = uniform(0, max_delay)
7           playNote(note, beats = next_delay, sustain = duration - offset)
8           offset += next_delay
9       fastForward(duration - offset)
```

This code is a little more complicated than the previous example, but let's talk through it one line at a time. On the first line, we're importing the **uniform** function. On line 2, we're defining our rolled function with two parameters: a list of notes in a chord and the number of total beats. On line 3, we're defining a constant value that defines the maximum value that our offset between two notes can take. Higher values will create a more spaced-out sound, and lower values will create more closed, tighter-sounding chords. On line 4, we initialize a variable to track the total offset at each step, which starts at zero. Starting on line 5, we iterate through each note of the chord. At each step, we calculate the offset to the next note and then call **playNote**. Finally, on line 9, we move the play-head the remaining number of beats.

If this code seems confusing, that's okay. We'll dive more into understanding this kind of code in later chapters. You can treat this function as another tool in your coding toolkit.

OPTION 4: Arpeggios

Another way of playing chords is to play one note at a time. This method is called an **arpeggio**. The order you play the notes doesn't matter. You can start at any note and play the notes of the chord in any order to get the sound that's right for your track. In the code below, we're starting with the lowest note and working up in increasing order of pitch.

```
1   def arpeggio(chord, total_beats):
2       duration = total_beats / len(chord)
3       for note in chord:
4           playNote(note, beats = duration)
```

On the first line, we set up our function definition, which takes two parameters: a list of notes in a chord and the total number of beats to play the chord. On the second line, we calculate the duration of each individual note by evenly dividing the total number of beats by the number of notes in the chord. On lines 3 and 4, we use a for loop to iterate through the list of notes in the chord, playing each note for the same number of beats.

OPTION 5: Patterned arpeggios

This arpeggio function is ok, but let's try to make something a little more interesting. Remember, we don't have to play each note evenly *and* we can switch up the order of the chord. Let's define a function that takes a seventh chord, which has exactly four notes, and play it over the duration of a measure. We're going to use the concept of **indexing**, which we'll read more about in the next chapter. For now, just know that indexing is how we access specific elements of a list. To index a list in Python, you use square brackets around the position of the element you want to use. These positions start at zero, so if we have a variable **chord** and we want to access the first element of the list, the root, we would type the following:

```
chord[0]
```

With this in mind, let's define a function. We have four beats to work with and four notes—that's indices 0 through 3. Here's a quick example of what's possible, but you can experiment with different note variations and note orders to get different effects.

```
1  def my_pattern(chord):
2      playNote(chord[0], 0.75)
3      playNote(chord[1], 0.25)
4      playNote(chord[2], 0.5)
5      playNote(chord[1], 0.5)
6      playNote(chord[3], 0.5)
7      playNote(chord[2], 0.5)
8      rest(1)
```

In this function, we aren't doing anything fancy to iterate through the chord, and we don't need to calculate our beats each time. Our beat values for lines 2 through 8 add up to exactly 4.0 beats. If you're following along, you can try tweaking these to be different values that still up to 4.0 beats. You can also try tweaking the indices of the notes to change which chord tone plays.

5 Scales, keys, and melody

Scales are patterns of notes played one at a time in ascending or descending order of pitch. Most scales span one octave using some subset of the 12 possible notes on the piano keyboard. When the scale completes the octave, the pattern starts over. **Keys** are similar to scales except that the ordering of notes doesn't matter, and they contain all of the notes in the scale regardless of the octave that you start on. Keys are like templates that help us select notes and chords that we know will sound good together. Keys give harmonic and melodic structure to music.

5.1 Chromatic scale

The building blocks of scales are half steps and whole steps. Half steps are the smallest interval commonly used in music and are the distance between two notes that are next to each other in pitch and on the piano keyboard. Whole steps are made up of two half steps.

The most basic scale is the chromatic scale. In this scale, every note is exactly one half step up from the previous note. This scale can start on any note and spans an octave in 12 notes. Starting with a C on the piano keyboard, we would have the following notes:

C C♯ D D♯ E F F♯ G G♯ A A♯ B

Or, using MIDI (Musical Instrument Digital Interface) note numbers we could also write:

48, 49, 50, 51, 52, 53, 54, 55, 56, 57, 58, 59

DOI: 10.4324/9781003033240-9

Playing a chromatic scale in TunePad is easy using a loop:

```
1   # loop from 48 up to (but not including) 60
2   for note in range(48, 60):
3       playNote(note)
```

If we wanted to play a chromatic scale starting on a different root note, we could just change the numbers in the **range** function above.

5.2 Major and minor scales

Perhaps the most important scales in Western music are the major and minor scales. These scales each use 7 out of the 12 possible notes in an octave. There are 12 major scales and 12 minor scales, 1 for each possible starting pitch. After the seventh note, the next note would be the first note—or **tonic**—an octave up. Scales are named by their tonic and quality in the same way that chords are named. A major scale starting on note D would be called *D Major*.

Major scales are commonly described as cheerful and happy (like major chords). The major scale is made up of the following intervals:

whole step, whole step, half step, whole step, whole step, whole step, half step

The major scale starting on C would have the notes shown in Figure 5.1.

In the MIDI version you can see that the whole steps skip up by two notes, while the half steps skip up by one step.

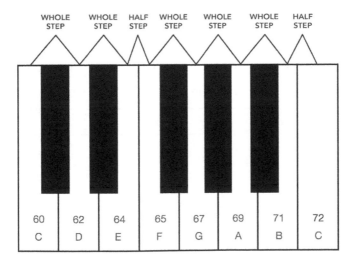

Figure 5.1 Whole step and half step intervals of the C major scale.

Note names	C	D	E	F	G	A	B	C
MIDI numbers	48	50	52	53	55	57	59	60
Intervals		WS	WS	HS	WS	WS	WS	HS

Minor scales also use seven notes out of each octave, but in a different order than major scales. This difference in intervals contributes to the different emotional connotation of the scale. Minor scales are commonly described as sad, melancholy, and distant. A minor scale starting on C would have the following notes:

Note names	C	D	E♭	F	G	A♭	B♭	C
MIDI numbers	48	50	51	53	55	56	58	60
Intervals		WS	HS	WS	WS	HS	WS	WS

Major and minor scales are both examples of **modes**. Modes are simply different ways of ordering the intervals of a scale.

5.3 Pentatonic scales

The pentatonic scales are a subset of the notes of the major and minor scales. There are five notes in a pentatonic scale. These scales have no half step intervals, which results in less dissonance between the notes. Many common melodies are based on pentatonic scales, especially in folk and pop music. The melody of *Amazing Grace* is pentatonic, as is Ed Sheeran's *Shape of You*.

There are both major and minor pentatonic scales. The major pentatonic is created by omitting the fourth and seventh notes of the major scale. The minor pentatonic omits the second and sixth notes of the minor scale. You can experiment with the sound of the pentatonic scale by playing only the black keys of the piano keyboard, which forms either an F♯ major pentatonic scale or a D♯ minor pentatonic scale (Figure 5.2).

The F♯ Major pentatonic scale uses only the black keys of the keyboard.
The D♯ Minor pentatonic scale starts with D♯ and uses only black keys as well.

C Major pentatonic scale

Note names	C	D	E	G	A	C
MIDI numbers	48	50	52	55	57	60
Intervals		WS	WS	m3	WS	m3

C Minor pentatonic scale

Note names	C	E♭	F	G	B♭	C
MIDI numbers	48	51	53	55	58	60
Intervals		m3	WS	WS	m3	WS

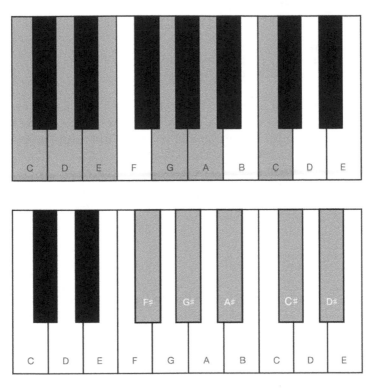

Figure 5.2 C Major Pentatonic Scale (top) and F♯ Major Pentatonic Scale (bottom).

5.4 Building scales in TunePad

Building scales in TunePad is similar to building chords. Because scales are just patterns of intervals (spaces between notes), we can create short functions to generate scales. Every major scale has an identical pattern of intervals, and the same is true for minor scales as well. The only thing that changes is the starting note. To generate scales in TunePad, all we need to do is decide what note to start on and then apply the pattern to this starting note.

One of the advantages of thinking about music in terms of computer code is that we don't have to memorize endless scales and combinations of notes and chords that make up different keys. Professional musicians train for years to learn how to play different scales without having to think about it so that they can fluidly switch from one key to another. This is part of what makes improvisational musicians so impressive. Playing a solo means knowing exactly which notes and chords can be played and how those notes and chords relate to a genre or theme of a piece being performed.

Here's a quick example of generating a scale with Python code in TunePad:

```
def majorScale(tonic):
    intervals = [ 0, 2, 4, 5, 7, 9, 11 ]
    return [ i + tonic for i in intervals ]
```

The example above uses a new Python concept called a **list comprehension**. A list comprehension is a shorthand way to create a list in Python. Line 2 uses a list comprehension to create a new list consisting of each element of the intervals list added to the tonic value:

```
[ i + tonic for i in intervals ]
```

This is equivalent to writing:

```
result = [ ]
for i in intervals:
    result.append(i + tonic)
```

The second version is a little more cumbersome to write than the first version using list comprehensions, although either version is fine to use.

5.4.1 Major scale

Intervals: [0, 2, 4, 5, 7, 9, 11]
Notation: C major, CMaj, CM, C
Python Function with List Comprehension:

```
1 | def majorScale(tonic):
2 |     intervals = [ 0, 2, 4, 5, 7, 9, 11 ]
3 |     return [ i + tonic for i in intervals ]
```

Here's an alternative Python function with a loop instead of a list comprehension:

```
1 | def majorScale(tonic):
2 |     intervals = [ 0, 2, 4, 5, 7, 9, 11 ]
3 |     scale = [ ]
4 |     for i in intervals:
5 |         scale.append(i + tonic)
6 |     return scale
```

Here's a third variation with no loop and no list comprehension:

```
1 | def majorScale(tonic):
2 |     return [ tonic, tonic + 2, tonic + 4, tonic + 5,
3 |              tonic + 7, tonic + 9, tonic + 11 ]
```

5.4.2 Minor scale

Intervals: [0, 2, 3, 5, 7, 8, 10]
Notation: C minor, Cmin, Cm
Python Function:

```
1   def minorScale(tonic):
2       intervals = [ 0, 2, 3, 5, 7, 8, 10 ]
3       return [ i + tonic for i in intervals ]
```

5.4.3 Major pentatonic scale

Intervals: [0, 2, 4, 7, 9]
Python Function:

```
1   def majorPentScale(tonic):
2       intervals = [ 0, 2, 4, 7, 9 ]
3       return [ i + tonic for i in intervals ]
```

5.4.4 Minor pentatonic scale

Intervals: [0, 3, 5, 7, 10]
Python Function:

```
1   def minorPentScale(tonic):
2       intervals = [ 0, 3, 5, 7, 10 ]
3       return [ i + tonic for i in intervals ]
```

5.4.5 Chromatic scale

Intervals: [0, 1, 2, 3, 4, 5, 6, 7, 8, 9, 10, 11, 12]
Python Function:

```
1   def minorPentScale(tonic):
2       intervals = [ 0, 3, 5, 7, 10 ]
        return [ i + tonic for i in intervals ]
```

Try these functions at https://tunepad.com/examples/build-scales.

5.5 Playing scales in TunePad

Now that we've seen how to build a scale, we can use the functions from the previous section to play music. Unlike with chords, the notes of a scale aren't usually played all at once. The most basic way to play a scale is to play one note at a time, in ascending order. Somehow we have to access each element of the list individually and pass that to **playNote** (Figure 5.3).

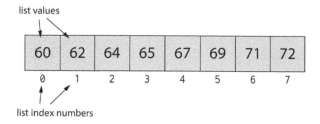

Figure 5.3 A representation of a list with values and indices.

Each element can be accessed using its position in the list—called an *index*. In coding, the first element of a list is at index 0; the second element of the list is at index 1; the third at index 2; and so on. In Python you can also access the last element of a list at index −1. Accessing individual elements of a list is referred to as *indexing*. In code, we do this by using square brackets and the index number. We can also use this technique to change the value of individual elements in a list.

```
1  notes = [ 60, 62, 64 ]
2  notes[2] = 66  # replace the value 64 with 66
3  playNote( notes[0] )
4  playNote( notes[1] )
5  playNote( notes[2] )
```

In line 1, we define a new list called **notes** with three values. In line 2, we replace the value at index 2 with the new value of 66. In lines 3 through 5, we play each note of the list, one at a time. **One of the most confusing parts about computer programming for beginners is that lists start at index 0 and end at an index one less than the length of the list.** However, with a little practice this becomes less and less confusing.

Note: If you try to index into a list with an index that doesn't exist, Python will stop running and complain with an **IndexError**. Because indices start at zero, valid indices are 0 all the way up to the length of the list minus 1.

Another way to iterate through a list is by using a **for loop**. Previously, when we've seen loops, we've used them to do the exact same operation multiple times in a row. Recall the syntax of a for loop:

```
for var in range(start, stop):
```

We can replace the **range(start, stop)** part of a for loop with a list instead. This will execute the body of the loop once for every element in the list. There's a special variable here called the "loop variable" that gets set to the value of each consecutive element in the list every time the loop repeats. In the code above, **var** is the loop variable, but you can use any valid Python variable name. For example, here's a loop that plays all the notes of a major scale starting on note 60.

```
for note in majorScale(60):
    playNote(note)
```

In this example, **note** is our loop variable. For every iteration of the loop it gets set to the next note in the scale. Give this a try at
https://tunepad.com/examples/play-scales.

5.6 Other kinds of scales

There are many other types of scales, but variants of the major and minor scale are the most common in popular music. Other scales that we don't cover here include the set of church modes, the whole tone scales, the diminished scales, and the modes of limited transposition.

Above we have discussed scales common to Western music, but the concept of collections of notes is cross-cultural. The Arabic maqam (or مقام) is the system of melodic modes in traditional Arabic music used in both compositions and improvisations. In Indian classical music the Raga are collections of melodic modes and motifs, each connoting a distinct personality or emotion. Gamelan music in Indonesia is organized by Pathet, which is a system of hierarchies of notes in which different notes have prominence. Composers from the West have often borrowed—or in some cases, stolen—these scales for their own music. This raises many issues of appropriation and exploitation within the music industry. The music industry has a long history of marginalizing groups while also profiting off of cultural traditions without properly compensating or recognizing musical provenance.

5.7 Keys

When writing music, there are seemingly endless notes to choose from. Keys are one way to narrow down the question of what note to choose. Keys are the underlying organizational framework of most music and encode both melodic and harmonic structures and rules. Knowing these rules (and how to break them) helps us to write music that listeners can easily comprehend and appreciate.

The concept of keys is closely related to that of scales. Keys are composed of all of the notes in all of the octaves that make up the scale with the same name.

For example, the notes in the key C major are the same as the notes in the C major scale across all octaves. But, while scales are usually played in increasing or decreasing order of pitch, the ordering of notes in a key doesn't matter. The notes that are part of a given key are called **diatonic**, and the remaining notes that are not part of that key are called **chromatic**.

Hundreds of years ago, different keys used to be associated with different emotions, so composers would choose specific keys that reinforce the mood of their composition. This is because the intervals in each key were slightly different due to the system of tuning; different keys were actually aurally distinct from one another. In modern times, each key is made up of the exact same intervals.

5.8 Circle of Fifths

Keys are organized according to the **Circle of Fifths**. The Circle of Fifths is essentially a pattern of intervals. Moving clockwise around the circle is moving the tonic note up by a fifth from the previous key. This adds one raised—or sharp—note as the seventh note of the scale. Alternatively, moving counterclockwise raises the tonic by a fourth and is often referred to as the Circle of Fourths. This adds one flat note as the fourth note of the scale (Figure 5.4).

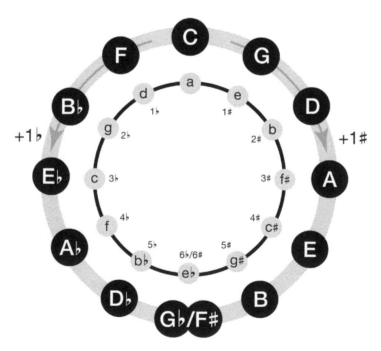

Figure 5.4 The Circle of Fifths arranges musical keys.

Major and minor keys that share all of the same notes are considered relative keys. For a minor scale, the relative major scale starts on the third note; for a major scale, the relative minor starts on the sixth note. The relative minor key of C major is A minor, and the relative major of A minor is C major.

Keys that are adjacent on the Circle of Fourths or Fifths—for example, D major and G major—share nearly all of the same notes and are considered *closely related*. The relative major or minor key for a given key is also considered closely related. Generally, when a song changes keys—also known as modulating—it goes to one of the closely related keys. Because closely related keys share most of the same notes, modulating to one of these keys is less jarring to the listener.

5.9 Melody

Melody is the central component of much of the music we listen to. It's the part of a song that gets stuck in your head. Much of what goes into great melody writing is intuition and practice, but knowing a bit of theory can help get you started. Melodies have two essential parameters: pitch and rhythm. These elements are of equal importance, but in this section, we'll mostly be looking at pitch.

When it comes to writing melodies, understanding that you are working within the confines of keys, scales, and a harmonic chord progression is a great place to start. We can use our song's harmony to provide a structural scaffold. Often, melodies place the chord tones from the harmony on the strong beats of the measure (beats one and three). These tones are consonant with the harmony, meaning that they sound pleasing. The simplest melody might stick solely to these chord tones. In the example below, we are only playing the chord tones of C major and D minor:

```
1   # over C major
2   playNote(55, 0.75)
3   playNote(55, 0.25)
4   playNote(52, 1)
5   playNote(48, 0.5)
6   rest(1.5)
7
8   # over D minor
9   playNote(57, 0.75)
10  playNote(57, 0.25)
11  playNote(53, 1)
12  playNote(50, 0.5)
13
```

Follow along with these examples at https://tunepad.com/examples/simple-melody.

Dissonance is also a powerful tool in melody writing. This can add interest and variation and sometimes have an intense emotional impact on listeners. A melody with no dissonance, that only plays the chord tones, becomes boring. One way to utilize dissonance is to add notes in between our chord tones to fill in our melody. We can choose the notes that correspond with the scale based on our song's key. In the example below, we are now filling in the space between the last two notes of each measure:

```
1   # over C major
2   playNote(55, 0.75)
3   playNote(55, 0.25)
4   playNote(52, 0.5)
5   playNote(50, 0.5) # passing tone
6   playNote(48, 0.5)
7   rest(1.5)
8
9   # over D minor
10  playNote(57, 0.75)
11  playNote(57, 0.25)
12  playNote(53, 0.5)
13  playNote(52, 0.5) # passing tone
14  playNote(50, 0.5)
```

Something to consider when writing a melody is **contour**. Contour describes the shape the melody takes: the natural rises and falls in pitch. A melody can either move stepwise to adjacent notes or leap to more distant notes. This motion can either decrease or increase in pitch. That means we have four types of motion a melody might take, each with different connotations. For instance, we might hear a melody that opens with a large leap as more emotional. You can expect most melodies to be within the range of about an octave to an octave and a half. In the example below, we'll combine the idea of leaps to chord tones and passing tones:

```
1   # over C major
2   playNote(55, 1)
3   playNote(64, 1.5) # large leap
4   playNote(60, 1.5)
5
6   # over D minor
7   playNote(57, 1)
8   playNote(65, 1.5) # large leap
9   playNote(67, 1) # passing tone
10  playNote(69, 0.5)
11
```

If we consider contour and pitch content as a vertical phenomenon, we can think of melodic form as a horizontal structure. We can break melodies into parts called **phrases**. If melodies are paragraphs, then phrases are like musical sentences. They are complete thoughts that are punctuated and combined to form more complete and cohesive ideas. Phrases are often 2, 4, or 8 bars in duration. These phrases are combined to form larger structures, which become the overall song form. We'll explore this more in Chapter 9.

The principles of repetition and variation work in opposition to one another. In writing melodies, there generally needs to be enough repetition so that a listener has something to latch onto. But with too much repetition, a melody becomes boring. A catchy melody is the result of striking a balance between these two forces.

One way to build intuition about melody writing is to analyze melodies from artists you like and want to emulate. The critical listening skills that you develop from analyzing existing melodies is directly applicable to writing your own melodies. Experimentation and improvisation are other great ways to build up this intuition. You can try tapping out rhythms to serve the basis of a melody, or play around on a piano or another instrument.

Try playing around with our automatic melody generator at
https://tunepad.com/examples/melody-gen.

Interlude 5
LEAN ON ME

Bill Withers

Let's practice using chords by recreating a small part of the piano harmony from the song *Lean on Me* by Bill Withers (1972), Columbia Records. Here's a simplified version of the chord structure that you can try entering into a TunePad project.

```
# Chord Variables
Cmaj = [ 48, 52, 55 ]
Dmin = [ 50, 53, 57 ]
Emin = [ 52, 55, 59 ]
Fmaj = [ 53, 57, 60 ]
Bdim = [ 47, 50, 53 ]
```

Chord	Beats	Python	Chord	Beats	Python
C major	4	playNote(Cmaj, beats = 4)	C major	4	playNote(Cmaj, beats = 4)
C major	1	playNote(Cmaj)	C major	1	playNote(Cmaj)
D minor	1	playNote(Dmin)	D minor	1	playNote(Dmin)
E minor	1	playNote(Emin)	E minor	1	playNote(Emin)
F major	4	playNote(Fmaj, beats = 4)	F major	4	playNote(Fmaj, beats = 4)

DOI: 10.4324/9781003033240–10

F major	1	`playNote(Fmaj)`	F major	1	`playNote(Fmaj)`
E minor	1	`playNote(Emin)`	E minor	1	`playNote(Emin)`
D minor	1	`playNote(Dmin)`	D minor	1	`playNote(Dmin)`
C major	4	`playNote(Cmaj, beats = 4)`	C major	4	`playNote(Cmaj, beats = 4)`
C major	1	`playNote(Cmaj)`	C major	1	`playNote(Cmaj)`
D minor	1	`playNote(Dmin)`	D minor	1	`playNote(Dmin)`
E minor	1	`playNote(Emin)`	E minor	1	`playNote(Emin)`
E minor	3	`playNote(Emin, beats = 3)`	B dim	3	`playNote(Bdim, beats = 3)`
D minor	4	`playNote(Dmin, beats = 4)`	C maj	4	`playNote(Cmaj, beats = 4)`

You can see the full code here: https://tunepad.com/interlude/chord-progressions. One thing to notice is how the chord progression mirrors the emotion of the song as a whole. Withers mixes the upbeat ("I'll be your friend. I'll help you carry on") with the harsh reality of life ("We all have pain. We all have sorrow"). The harmony starts on a major chord (C major) but then passes through a succession of minor chords (D minor, E minor) before eventually landing on more encouraging major chords for prolonged notes (F major). It's as if the harmony is also saying that we're going to go through some hard times, but it'll all work out in the end.

The version above is slightly modified from the original in that we're using simplified chords and a diminished B chord at the end that slides into a C major. As you listen to it, notice how the B diminished feels unstable as if it needs to resolve into the C major to bring the harmony full circle to signal a transition in the song.

More elegant code

One tempting way to code this up would be to just type all of the **playNote** functions, one after another. This works, but it's not necessarily the most elegant way to express the music. When you're coding, there's always more than one way to solve a problem, so it's good to get into the habit of asking if there are other, easier ways to accomplish things. For example, what if we wanted to change the velocity of all of the chords? We'd have to edit one line at a time or use find and replace. As an alternative, what if we put all of the chords into a list and then iterated through that list with a for loop?

```
chords = [ Cmaj, Cmaj, Dmin, Emin, Fmaj, Fmaj, Emin, Dmin, Cmaj ]
for chord in chords:
    playNote(chord)
```

This would be an improvement. If nothing else, we would have reduced the number of lines needed to play the harmony. The obvious problem is that it won't work because the notes are different lengths. Some are long (four beats) and others are short (one beat). But this code plays all the chords with equal duration.

If there were an easy way to iterate through two lists at the same time, we could make one list with the chords and another with the durations. Python includes exactly this kind of feature with something called the **zip** function. Think of it like a zipper that merges two Python lists together instead of two pieces of fabric. It walks through the lists, element by element, and merges them together into pairs of values. The result looks something like this:

```
chords = [ Cmaj, Cmaj, Dmin, Emin, Fmaj, Fmaj, Emin, Dmin, Cmaj ]
durations = [ 4, 1, 1, 1, 4, 1, 1, 1, 4 ]
for chord, duration in zip(chords, durations):
    playNote(chord, beats = duration)
```

Here's the complete example or you can try it online: https://tunepad.com/interlude/lean-on-me.

```
1    CM = [ 48, 52, 55, 55 + 12 ]
2    Dm = [ 50, 53, 57, 57 + 12 ]
3    Em = [ 52, 55, 59, 59 + 12 ]
4    FM = [ 53, 57, 60, 60 + 12 ]
5    Bd = [ 47, 50, 53, 53 + 12 ]
6
7    chords = [ CM, CM, Dm, Em, FM, FM, Em, Dm, CM, CM, Dm, Em ]
8    durations = [ 4, 1, 1, 1, 4, 1, 1, 1, 4, 1, 1, 1, 3, 4 ]
9
10   for chord, duration in zip(chords + [ Em, Dm ], durations):
11       playNote(chord, beats = duration)
12
13   for chord, duration in zip(chords + [ Bd, CM ], durations):
14       playNote(chord, beats = duration)
```

6 Diatonic chords and chord progressions

Now that we have some familiarity with chords, the question is how to use them. How can we reduce hundreds of chords and thousands of combinations of chords down to a manageable set of options? How can we explore the creative musical space that chords provide without feeling overwhelmed?

One answer to these questions is to use keys to select subsets of chords that we know will sound good together. From there we can follow guidelines for arranging chords into sequential patterns called *progressions* that will support the various harmonic elements that come together in a piece of music.

This chapter introduces the traditional **Roman numeral** system for referring to chords that fit with a particular key along with methods for choosing chord progressions. Along the way we'll code functions for creating chord progressions in any key using the concept of Python *dictionaries*.

6.1 Diatonic chords

A **diatonic** chord is any chord that can be played using only the seven notes of the current key. For example, if you're working in the key of C major, the diatonic chords consist of all of the chords you can play with only the white keys on the piano keyboard:

$$C \quad D \quad E \quad F \quad G \quad A \quad B$$

Here are the main diatonic chords we can make with just these seven notes (Figure 6.1):

DOI: 10.4324/9781003033240-11

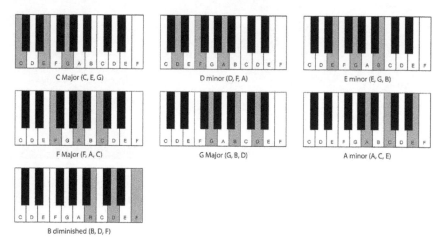

Figure 6.1 The seven diatonic chords of C major.

Hear these chords at https://tunepad.com/examples/diatonic-chords.

If you have a piano keyboard handy, try playing with these seven chords to get a feel for how they sound. What emotions do you feel as the chords ring out? What patterns of chords sound good together? No matter what key we're in, there will always be seven diatonic chords, one for each note in the scale. To build a diatonic chord, just pick any note from the scale as the root of the chord. Then go up two notes for the "third" of the chord, and up two notes again for the "fifth" of the chord. You can keep moving up by two notes of the scale to get the "seventh" and "ninth" and so on.

We can refer to these chords by their scale degree (first, second, third, fourth, fifth, sixth, and seventh). A chord built on the fifth note of a scale would be the "five" chord for that key. If we're in any *major key*, the first, fourth, and fifth diatonic chords will have a major quality (regardless of the starting note of the key). Further, the second, third, and sixth chords will always have a minor quality, and the seventh chord is diminished. For example, in C major, we would have the following diatonic chords:

CMaj   Dmin   Emin   FMaj   GMaj   Amin   Bdim

This pattern of chord qualities is the same for any major key because all of the major keys have the same pattern of note intervals. The same idea is true for minor keys. Because all minor keys have the same pattern of intervals, the quality of chords stays consistent. The first, fourth, and fifth diatonic chords are minor quality. The third, sixth, and seventh chords are major quality. The second chord is diminished. In C minor, for example, we would have the following diatonic chords:

Cmin   Ddim   E♭Maj   Fmin   Gmin   A♭Maj   B♭Maj

In popular music, chord progressions are made up almost entirely of diatonic chords. Recall from the previous chapter that the melody of a song is also built on this harmonic scaffold. Melodies often use primarily the notes from the underlying chord progression, because these notes are more consonant and pleasing. Notes from outside the harmony are typically used in passing as decoration or as a neighboring tone.

6.2 Roman numerals

Each of the 12 major keys and the 12 minor keys have 7 diatonic chords, giving us an overwhelming total of 168 diatonic chords. To reduce this complexity, musicians, producers, and composers use a system of **Roman numerals** to refer to different diatonic chords by their scale degree rather than their specific name. If the chord has a major quality, it gets an uppercase Roman numeral. If it has a minor quality, it gets a lowercase Roman numeral. Each key also has one diminished chord, which is both lowercase and has an accompanied ° symbol.

> **Note**: Roman numerals are a system of numbering which originated in ancient Rome. In this system, numbers are composed of combinations of letters. In music, only the numerals corresponding to numbers one through seven are used: I, II, III, IV, V, VI, and VII.

That gives us the following Roman numerals for all of the diatonic chords of any key.

Scale degree	1st	2nd	3rd	4th	5th	6th	7th
Major keys	I	ii	iii	IV	V	vi	vii°
Minor keys	i	ii°	III	iv	v	VI	VII

With this system there are just these seven symbols to focus on instead of 168. Roman numerals take some getting used to, but they give us a language for thinking about chord progressions without having to refer to the name of each specific chord. In coding or mathematics, this kind of generalization is called an **abstraction**. Abstraction means removing the specific details of individual situations and focusing instead on the bigger picture patterns. In coding, we use language constructs like variables, functions, and parameters to create abstractions that reduce the complexity of our code and the problems we're trying to solve.

6.3 Tendency tones and harmonic functions

Now that we know how to refer to diatonic chords by their names, how do we order them into pleasing chord progressions? The ordering of chords within a progression isn't random. The notes that make up chords have different tendencies relative to one another—meaning that the listener hears them as wanting to resolve in expected ways when played with other notes in a chord progression. The strongest of these tendencies is the pull of the seventh note of a scale back to the root note of the scale (the tonic). In most cases, we hear this note as wanting to resolve upward by a half step back to the tonic. If this note does not resolve, we often hear the progression as incomplete. The second and fifth notes of a scale also have a strong pull back to the tonic.

The tendencies of the individual notes give each chord a characteristic or *function*. You can think of a chord's function as its desire—how it relates to the previous chords and how it *wants* to move the music forward. Chords are described as having three primary functions: **tonic**, **predominant**, and **dominant**. Most chord progressions typically progress from tonic to predominant to dominant back to tonic.

Tonic chords	Predominant chords	Dominant chords
I, iii, vi	ii, IV	V, vii°
1st, 3rd, 6th	2nd, 4th	5th, 7th

Chords that have the same function also share many of the same notes. For example, iii (3) chords and vi (6) chords share two of the same notes with the tonic chords (I), so they're grouped together. The predominant ii and IV chords also share two notes in common, as do the dominant V and vii° chords (Figure 6.2).

Chord functions might best be described as rough guidelines that help inform our decisions about composition. There are also many variations to the basic chords. Chords can be inverted (meaning that the root is no longer the lowest note), extended with additional notes to add color, or combined with "chromatic" chords that include notes from outside of the main key. Secondary dominant chords are dominant chords borrowed from other related keys. Knowing which chord to use in any given context comes down to musical experience, taste, and the conventions of different genres.

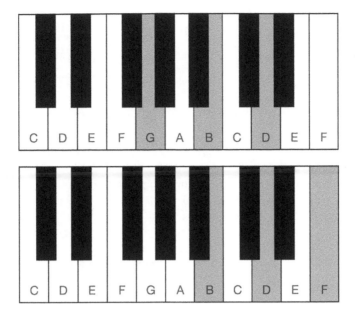

Figure 6.2 The dominant V and vii° chords share two notes in common. This is easy to see when you line piano diagrams up vertically.

6.4 Chord progressions

Many common chord progressions follow the scheme of **tonic → predominant → dominant**. The tonic brings stability and grounding. The predominant is a departure from this stability that builds tension. The predominant pulls toward the dominant, which eventually resolves back to the tonic. After a chord progression finishes, it starts over. Different genres of music have different harmonic rules and standard chord progressions, but here are flow charts that help visualize common chord progression patterns (Figure 6.3).

Here's an example of how to use these charts. If we start with the tonic I chord on top, we might move down to the dominant V chord, and then slide up to the tonic prolongation vi chord—a subtle tease with resolution. We could then go to the predominant IV chord before the progression resolves back to the tonic I chord and repeats. This progression would be I → V → vi → IV (1, 5, 6, 4), which is an extremely common pattern in popular music (Figure 6.4).

Contemporary pop music uses many of the same progressions as early Rock and Roll and Blues music—much of early rock music came out of the Blues. As a result, many early rock songs are built on Blues progressions, most notably I-IV-V. One of the most ubiquitous progressions—especially in early rock—is the "doo wop" progression I-vi-IV-V. The same chords can be reordered to form our I-V-vi-IV example.

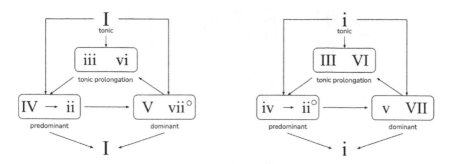

Figure 6.3 Flowcharts for generating chord progressions in major and minor keys.

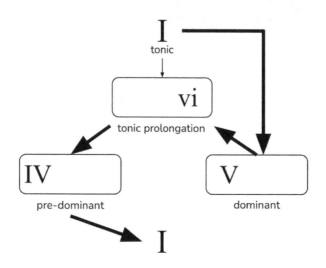

Figure 6.4 Example of using the flowchart to generate a chord progression.

Common major progressions	Common minor progressions
I - V - vi - IV	i - VI - VII
I - IV - V	i - III - iv - v
I - V - vi - iii - IV	i - VI - III - VII
I - vi - ii - V	i - v - VI
I - vi - IV - V	i - iv - v
I - iii - IV - V	VI - iv - v — i

Hip-hop songs center more around rhythm and vocals and tend to use shorter progressions, often in a minor key with only one or two chords such

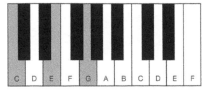

C Major in root position

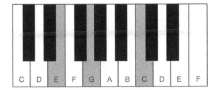

1st inversion of C Major

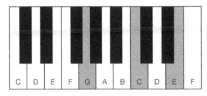

2nd inversion of C Major

Figure 6.5 C major chord in root position (top), first inversion (bottom left), and second inversion (bottom right).

as i–V, i–VI, and i–ii°. Hip-hop developed alongside advances in recording technology that allowed early artists to remix samples from other songs, and, as a result, the genre also borrows progressions from pop and rock music.

When writing chord progressions, one tactic is to borrow from existing songs to help you develop your own ear and begin to think critically about harmony. You can also experiment on your own. Use the harmonic conventions to narrow down some of the options, but also try breaking the rules as you become more confident.

6.5 Chord inversions

An **inverted** chord is just like an ordinary chord except that the root note is no longer the lowest pitch. Take C major as an example. When the root note, C, is also the lowest note of the chord, we say that the chord is in root position (Figure 6.5).

When the third of the chord is the lowest note, the chord is in its first inversion. In the case of C major, that means that E is now the lowest note. When the fifth of the chord is lowest, it's the second inversion, and so on. Each inversion has exactly the same notes as the root chord, but the ordering of the notes by their pitch is different.

6.6 Voice leading

Voice leading deals with the relationship between notes in consecutive chords in a progression. The principle behind voice leading is to treat

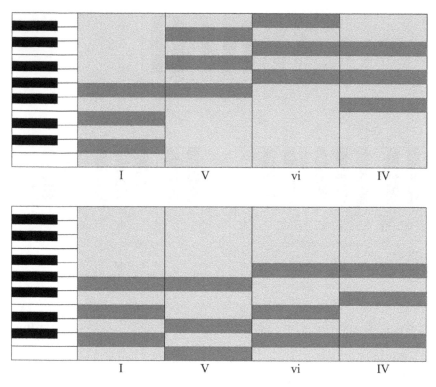

Figure 6.6 Chord progression I-V-vi-IV without voice leading (top) and with voice leading (bottom). Chords V, vi, and IV are inverted to reduce the pitch range and to minimize the movement of individual voices "singing" the notes of the chords.

each note of a chord as an individual melodic voice. Imagine three human vocalists each singing one individual note of a chord. Because we're considering each voice independently, the idea is to minimize the leaps one person's voice has to make between chords so that the progression is smoother and easier to sing. By considering the different possible inversions of each chord we can create more of a dovetailing effect with subtle shifts between successive chords. Not only will this improve the sound of your progressions, but it will also improve the potential playability of the music on instruments like the guitar, piano, or vocal harmony. The two figures below show the same progression with and without voice leading (Figure 6.6).

Hear these examples at https://tunepad.com/examples/voice-leading.

6.7 Python dictionaries

In Python, **dictionaries** or **maps** are unordered sets of data consisting of values referenced by *keys*. These keys aren't the same as musical keys. They're more like the kind of keys that open locked doors. Each different key opens its own door.

Dictionaries are extremely useful in programming because they provide an easy way to store multiple data elements by name. For example, if we wanted to store information for a music streaming service, you might need to save the song name, artist, release date, genre, record label, song length, and album artwork. A dictionary gives you an easy method for storing all of these elements in a single data object.

```
track_info = {
    "artist" : "Herbie Hancock",
    "album" : "Head Hunters",
    "label" : "Columbia Records",
    "genre" : "Jazz-Funk",
    "year" : 1973,
    "track" : "Chameleon",
    "length" : 15.75 }
```

Dictionaries are defined using curly braces with keys and values separated by a colon. Different entries are separated by commas. After defining a dictionary, we can change existing values or add new values using the associated keys. Similar to the way we access values in a list with an index, we use square brackets and a key to access elements in a dictionary.

```
track_info["artwork"] = "https://images.ssl-images-amz.com/
images/81KRhL.jpg"
```

In this line, because the key "artwork" hasn't been used in the dictionary yet, it creates a new key–value pair. If "artwork" had been added already, it would change the existing value. One thing to notice is that values in a dictionary can be any data type including strings, numbers, lists, or even other dictionaries. Dictionary keys can also be strings or numerical values, but they must be unique for each value stored.

6.8 Programming with diatonic chords

With Python code, there are many different ways to determine the diatonic chords for a given key. Here are the **majorChord**, **minorChord**, and **diminishedChord** functions from Chapter 4 again.

```
def majorChord(root):
    return [root, root + 4, root + 7]

def minorChord(root):
    return [root, root + 3, root + 7]

def dimChord(root):
    return [root, root + 3, root + 6]
```

We can use these functions to define variables for each diatonic chord in C major:

```
I   =   majorChord(48)
ii  =   minorChord(50)
iii =   minorChord(52)
IV  =   majorChord(53)
V   =   majorChord(55)
vi  =   minorChord(57)
viiO = diminishedChord(59)
```

This code is clear and readable, but it's not as reusable as it could be. What if we want to play in a different key? Or in a different octave? We'd have to change *each* line of code. As an alternative, we could write a function that takes the tonic as an input and returns a dictionary that maps Roman numerals to individual diatonic chords.

```
def buildChords(tonic):
    numerals_lookup = { "I" : majorChord(tonic),
                        "ii" : minorChord(tonic+2),
                        "iii" : minorChord(tonic+4),
                        "IV" : majorChord(tonic+5),
                        "V" : majorChord(tonic+7),
                        "vi" : minorChord(tonic+9),
                        "viiO" : diminishedChord(tonic+11)}
    return numerals_lookup
```

The method above works for major keys, but what if we wanted it to work with minor keys as well? We can add another parameter and an **if-else** statement to handle this as well.

```
def buildChords(tonic, mode):
    if mode == "major":
        numerals_lookup = {"I" : majorChord(tonic),
                           "ii" : minorChord(tonic+2),
                           "iii" : minorChord(tonic+4),
                           "IV" : majorChord(tonic+5),
                           "V" : majorChord(tonic+7),
                           "vi" : minorChord(tonic+9),
                           "viiO" : diminishedChord(tonic+11)}
```

```
    else:
        numerals_lookup = {"i"  : minorChord(tonic),
                           "ii0" : diminishedChord(tonic+2),
                           "III" : majorChord(tonic+3),
                           "iv"  : minorChord(tonic+5),
                           "v"   : minorChord(tonic+7),
                           "VI"  : majorChord(tonic+8),
                           "VII" : majorChord(tonic+10)}
    return numerals_lookup
```

These are far from the only solution for creating the diatonic chords for different keys. In general, there are almost an endless number of ways to solve complex problems in programming. Figuring out which approach is best for a given circumstance takes practice and experience, but your goal is usually to write code that is as simple and easy to understand as possible.

Try this code at https://tunepad.com/examples/chord-dictionary.

Interlude 6

RANDOM CHORD PROGRESSIONS

Here's a short Python example that generates and then plays random chord progressions using the charts in Figure 6.3. We can start with a table that maps each chord to a simplified set of possible transition chords. The table below on the left uses Roman numerals, and the table on the right uses Arabic numbers to show the same thing. Note that these tables don't include all of the possibilities from the flow charts above, but most of the possible transitions are included.

I	→	iii, IV, V, vi		1	→	3, 4, 5, 6
ii	→	I, V		2	→	1, 5
iii	→	IV		3	→	4
IV	→	I, ii, V		4	→	1, 2, 5
V	→	I, vi		5	→	1, 6
vi	→	ii, iii, IV		6	→	2, 3, 4

Now we can turn this transition table into a computer algorithm using Python.

DOI: 10.4324/9781003033240-12

STEP 1: Random chord algorithm

Create a new piano cell in a TunePad project and add this code.

```
1    from random import choice       # import the choice function
2
3    progression = [ 1 ]             # create a list with just one chord
4    chord = choice([3, 4, 5, 6])    # choose a random next chord
5
6    while chord != 1:               # repeat while chord is not equal to 1
7        progression.append(chord)   # add the next chord to the list
8        if chord == 2:              # if the current chord is 2
9            chord = choice([1, 5])  # then choose a random next chord
10       elif chord == 3:            # else if the current chord is 3
11           chord = 4               # …
12       elif chord == 4:
13           chord = choice([1, 2, 5])
14       elif chord == 5:
15           chord = choice([1, 6])
16       else:                       # the chord is 6
17           chord = choice([ 2, 3, 4 ])
18   print(progression)
19
```

There's a lot going on with this code, but let's break it down line by line.

Line 1 imports a function called **choice** from Python's **random** module. The **choice** function selects one element from a list at random. You can think of it as picking a random card from a deck.

Line 3 creates a variable called **progression** that consists of a list with only one element in it. This list will hold our finished chord progression, and we start it with the tonic chord, 1.

Line 4 picks the next chord at random. We use our transition table to select from 3, 4, 5, and 6 as possible next chords in the sequence. We save the random choice in a variable called **chord**.

Line 6 is a new kind of Python loop called a **while loop** that we haven't seen before. This loop repeats indefinitely until a certain condition is met. In our case, we're going to repeat the loop until our **chord** variable is equal to 1.

Line 7 is part of the while loop. It adds our new **chord** to the end of the **progression** list using the append function. The first time through the loop, the **progression** list will have two elements, 1 and whatever random chord was selected on line 4. Each additional time through the loop, line 7 will add another chord to the list.

Line 8 asks *if* our random chord is equal to 2. If so, it selects a random next chord based on the values in our transition table (on line 9).

Line 10 only gets used if line 8 is False. The **elif** is short "else if". So, this line says: otherwise, if the value of **chord** is 3, then set the next chord to 4.

Lines 12, 14, and 16 handle the next set of options for the value of **chord**, following the transition table.

Once the loop completes, **line 19** prints out the result. A sample output might be **[1, 5, 6, 3, 4, 2]**, but since this uses random selection, the output is likely to be different each time the code runs.

STEP 2: Play the chords

So how do our random chord progressions sound? We can add a few more lines of code at the end to play our progression in TunePad. Let's start by defining our diatonic chords in a dictionary. Instead of using the Roman numerals as keys, we're going to use the chord numbers.

```
tonic = 48
chords = {
    1 : majorChord(tonic),
    2 : minorChord(tonic + 2),
    3 : minorChord(tonic + 4),
    4 : majorChord(tonic + 5),
    5 : majorChord(tonic + 7),
    6 : minorChord(tonic + 9),
    7 : diminishedChord(tonic + 11) }
```

Next we can iterate over our **progression** list, playing each chord in turn.

```
for chord in progression:
    playNote(chords[chord], beats = 2)
```

You can try this code on TunePad here: https://tunepad.com/interlude/random-chords.

7 Frequency, fourier, and filters

Chapter 3 introduced the idea that different instruments and voices naturally fall into different ranges of the frequency spectrum, from low sounds like a bass to high sounds like hi-hats. In this chapter, we further explore the frequency spectrum, this time with an emphasis on techniques for mixing multiple layers of a musical composition into a cohesive whole. We show how sound can be decomposed into its component frequencies and how we can use filters and other tools to shape sonic parameters such as frequency, loudness, and stereo balance. We'll also show how to apply these standard filters and effects in TunePad using Python code.

7.1 Timbre

All sound is made up of waves of air pressure that travel outward from a sound's source until they eventually reach our inner ears. Sound waves that vibrate regularly, or periodically, are special kinds of audio signals which the human brain interprets as musical pitch. The rate of the vibration (or frequency) determines how high or low the pitch sounds. One surprising thing about musical notes is that they are almost never composed of just one frequency of sound. In fact, what people hear as one musical note is actually a whole range of frequencies stacked on top of one another. As an example, Figure 7.1 shows the sound energy generated by a flute playing a single note. The figure shows the energy level at different frequencies across the whole range of human hearing (from about 20 Hz

DOI: 10.4324/9781003033240-13

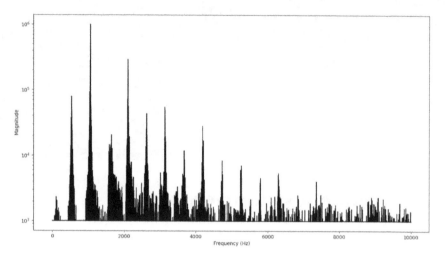

Figure 7.1 Sound energy generated by a flute playing a single note. The sound contains a series of spikes at regular "harmonic" frequency intervals.

to 20,000 Hz). The frequency level is shown on the horizontal axis and the energy level is shown on the vertical axis. The spikes in the graph show sounds generated by the flute at different frequency levels. So even though we only hear a single note, there are actually a whole range of frequencies present in the sound, one for each spike.

Frequency combinations like this allow people to distinguish different kinds of instruments from one another. It's how your brain can tell the difference between a trombone and cello, even when they're playing the exact same note. The many frequencies of a single sound are called its frequency spectrum (Figure 7.1, for example), and they create what is called timbre (pronounced "TAM-ber")—often called tone color or tone quality. Timbre is like the fingerprint of a sound.

Timbre	The unique fingerprint of a sound that results from how we perceive multiple frequencies combining together.
Fundamental frequency	The lowest (and usually) loudest frequency that we perceive as the pitch of a note.
Partials or overtones	Other frequencies beyond the fundamental frequency that are also present when we hear a note.
Harmonic frequency	Any frequency that is close to an integer multiple of the fundamental frequency.
Inharmonic frequency	Any partial that is not an integer multiple of the fundamental frequency.

What we perceive as the pitch of a note is usually the lowest of the frequencies present. This is known as the **fundamental frequency**, or often simply the **fundamental**. It's also usually the loudest of the frequencies. The remaining frequencies are referred to as **partials** or **overtones**. If the frequency of the partial is close to an integer multiple of the fundamental, then the partial is considered to be a **harmonic** of the fundamental. Otherwise, the partial is considered **inharmonic**. Most pitched or melodic instruments—such as saxophones, flutes, and guitars—have very harmonic spectrums (Figure 7.1). Non-pitched or percussive instruments often have very inharmonic spectrums, which means you don't really perceive the pitch of these instruments. You can hear how this sounds here: https://tunepad.com/examples/spectrums.

To help make this more clear, consider a sound consisting of the following frequencies: 200 Hz, 400 Hz, and 500 Hz (Figure 7.2, left). The fundamental of the sound would be 200 Hz, because it's the lowest (and loudest) frequency. The 400 Hz frequency would be the first partial and would be considered a harmonic because it's an integer multiple of the fundamental (400 Hz/200 Hz = 2). The 500 Hz frequency would be the second partial, but it's not a harmonic because 500 Hz/200 Hz = 2.5. In Figure 7.2 (right) we add 100 Hz as the new fundamental frequency. In this case, 200 Hz, 400 Hz, and 500 Hz would all be considered harmonics of 100 Hz because they are all simple integer ratios of 100 Hz (2, 4, and 5).

Listen to an example here: https://tunepad.com/examples/timbre.

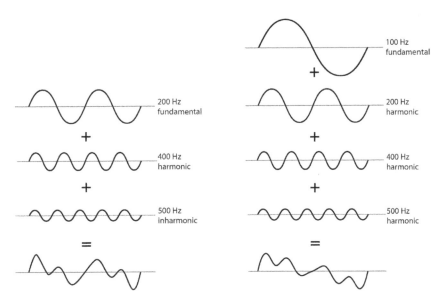

Figure 7.2 Frequency combinations: fundamentals, partials, harmonic, and inharmonic.

Almost all sounds consist of a complex combination of frequencies. The one exception to this rule is the sine wave (Figure 7.2 shows combinations of sine waves). Sine waves are made up of just one frequency with no other partials, and they are often described as sounding clear or pure because of this. Sine waves are easy to generate using electronics or a computer, but they rarely occur in nature, which means that they can also sound artificial and harsh.

It turns out that any periodic sound can be described as a combination of a (possibly infinite) number of sine waves forming the partial frequencies. Sounds that have very few partials, like a whistle, are often very close to sine waves. Sounds that have many partials, like a saxophone, have much richer and more complex waveforms. One way to imagine this is that sine waves are like the primary colors of paint that you can mix together to form every other color. Figure 7.3 shows how multiple sine waves at different harmonic frequencies can combine to approximate a more complex signal like a square wave.

7.2 Envelopes

There are other complex properties of sound waves that contribute to an instrument's timbre. One of the most important of these is how the volume of a sound evolves and changes over the duration of a note. This is called the sound's **envelope**. A simplified envelope is commonly described using four stages: Attack, Decay, Sustain, and Release or **ADSR** for short (Figure 7.4).

The ADSR envelope has both time components and amplitude (loudness) components. When you play a note on the piano or another instrument, the **attack** is the time from when the key is first pressed to when the note reaches its maximum volume. The **decay** is the time it takes the note to reach a lower secondary volume. The **sustain** is the loudness of this second volume. Finally, the **release** is how long it takes for the note to completely fade out. So, the attack, decay, and release are all measures of time, while sustain is a measure of loudness.

A sound like a snare drum has a sharp attack and a quick release, while sounds like cymbals or chimes have fast attacks but slower releases that ring out over longer periods of time. Other sounds like violins have both slower attacks and releases. The attack, decay, and release sections of an envelope can also be curved instead of straight lines, which sometimes better approximates the sound of real musical instruments. But it's important to remember that ADSR envelopes are always simplifications of reality. For example, the sustain of a piano note actually gradually decreases in volume over time until the note is finally released. We'll revisit the idea of ADSR envelopes in Chapter 10 to see how this can be applied when creating synthesized musical instruments.

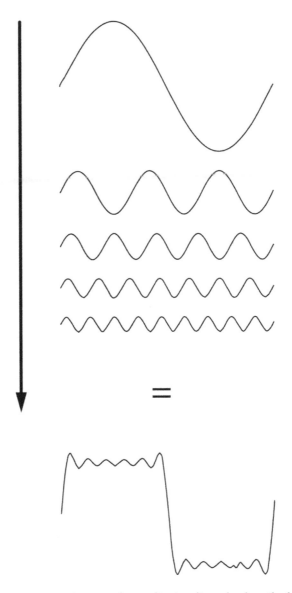

Figure 7.3 A square wave (or any other audio signal) can be described as a series of sine waves making up the partial frequencies.

7.3 Fourier

Jean–Baptiste Joseph Fourier was a French mathematician and physicist whose work in the 19th century led to what we now call **Fourier Analysis**, a process through which we can decompose a complex sound signal into its constituent individual frequencies. The idea is that we can take

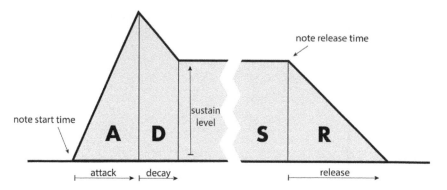

Figure 7.4 ADSR envelope.

any complex sound and determine all of the frequencies that contribute to the energy in the signal—basically finding a set of sine waves that can be combined to represent a more complex waveform. The composition of a sound signal by its frequency components is called the signal's **spectrum**, and it can be generated through a mathematical operation called the Fourier Transformation, which is an essential part of all modern music production. For any given slice of time, the spectrum might look like Figure 7.1. But we can also spread this information out over many time slices to visualize the frequency and amplitude of a signal as it changes over longer periods of time. This visualization is called a **spectrogram** (Figure 7.5). A spectrogram typically shows time on the horizontal axis, frequency on the vertical axis, and intensity of the different frequencies using heatmap colors. Warmer colors indicate more energy, while cooler colors indicate less energy.

This representation helps producers see and understand properties of sounds such as timbre and loudness. A spectrogram might show unwanted noise in the background, or point out that the audio is heavy on lower frequencies and sounds muddy, or that the audio is heavy on upper frequencies and sounds tinny. Through years of training, music producers can interpret spectrograms to visually understand how the various frequency bands contribute to a mix.

7.4 Mixing and mastering

Recording all the parts of a song is only one part of the process of creating a finished piece of music that's ready to be shared with the world. A music producer still has the task of making all of the various sonic layers work together as a cohesive whole. How does the bass line complement the rhythm? Does it interfere with the percussion sounds? Are the vocals getting drowned out by the instrumentals? Are the instruments competing

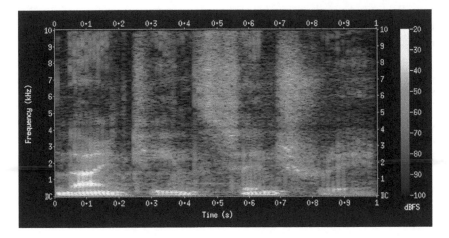

Figure 7.5 A spectrogram shows the intensity of frequencies in an audio signal over time. The heatmap colors correspond to intensity or energy at different frequencies. Time is represented on the horizontal axis and frequency in kilohertz on the vertical axis.

with one another? Is the overall mix too muddy or harsh or boomy? The process of **mixing** is about the overall compositional structure of a song and finding balance between the individual musical elements that have been recorded, sampled, or generated. Of course, *rough* mixes get put together throughout the creation process as the different parts of a song are recorded. For example, a recording studio would need a rough mix of the drums, bass, and keyboards before overdubbing the vocals. But the final mix is when all of the elements are balanced, placed in space, and blended together to make an artistic statement. Mixing can be a complex process that involves planning, deep listening, and a lot of patience to get it right.

Deep listening is the process of paying close attention to the relationship between musical elements in a song. This involves simultaneously being aware of the compositional structure and the frequency bandwidth of the individual sounds. With deep listening, you are paying attention to the different components and how they relate to each other. Is there a call-and-response between the guitar and the trumpet? Are they playing together? Should you separate them by adjusting the equalization to bring one out in the foreground or will a simple change in volume do the trick? Perhaps putting them in separate spaces within the stereo spectrum will work, or using a 100 ms delay effect to place it in lateral space away from the listener. There is a *lot* to experiment with, and it takes time to perfect the art of mixing. It takes practice to develop this listening skill, but with practice you will become keenly aware of the nuances. For instance, you'll hear when instruments with the same frequency bandwidth overlap. With digital tools that allow real-time frequency spectrum analysis, you can actually see where they overlap. Don't worry about getting it right the first time; mixing is a process that can require several iterations before you reach that "sweet" spot.

In this chapter, we're going to refer to each individual layer of music as a **track**.[1] It's common practice to record vocals, drums, bass, and so on, on separate tracks and then mix them together to form the final product. In TunePad, tracks are created using cells that can be assembled on a timeline as parts of a song. On a traditional mixing board, each **track** is a multifaceted tool used to shape sound elements in order to blend them cohesively with other elements in the song (Figure 7.6).

A few of the most important audio parameters to consider when mixing are panning, frequency manipulation (equalization), and gain (the loudness of each track). This is also where you can apply audio effects like reverb, echo, or chorus (some of which we'll cover in the next chapter).

7.4.1 Mixing tools

Before the widespread availability of digital production tools, recording studios used multi-track magnetic tape to record multiple elements of a song such as bass, drums, guitar, keyboards, and vocals. Large mixing consoles were then used to record and play back to each element on the tape (Figure 7.6). Studio infrastructure would route the signals from the individual tracks to and from the mixing board. Today this is mainly done using software and visualization tools that allow for more flexibility and precision.

Figure 7.6 Mixing console with magnetic tape.
Figure 7.6 is a photograph by Rebecca Wilson (flickr.com/people/34048699@N07). Creative Commons License creativecommons.org/licenses/by/2.0.

7.4.2 Panning and stereo

Most of the music that you listen to has multiple *channels* of audio data. Stereophonic, or **stereo**, recordings use the two different channels (left and right) to recreate the spatial experience of listening to music in natural acoustic environments.[2] In other words, when you're listening to music with headphones or earbuds, what you hear in your left ear is subtly (or not so subtly) different from what you hear in your right ear. Try removing one of your earbuds the next time you're listening to music to see if you can hear the difference. When you're experiencing live music, you have a physical position in space relative to the various musicians, vocalists, and other audio sources in the room. Your two ears are also pointed in opposite directions, meaning they receive different versions of the same audio scene. Music producers use the stereo spectrum to recreate this experience. In general, humans have evolved to process sound from these two sources to create a mental map of the physical space surrounding us. Think about a truck that drives past you. Even if you can't see the truck, your brain is able to tell you where the truck was and roughly how fast it was going based on frequency, loudness, and phase differences from your left and right ears.

The **panning** of a track refers to its position in this stereo spectrum. In practice, this means how much of the track comes out of the left and right speakers. Producers can create more depth to a song and replicate live recordings by controlling the panning of tracks. You can almost think of it like arranging musicians on a stage in front of a live audience. Humans are also better at perceiving the directionality of sound at higher frequencies, meaning we can easily tell which direction a high-pitched hi-hat sound is coming from, but we have a hard time telling which direction a bass line is coming from. As a result, producers will often pan higher-pitched sounds to the left or right, while leaving lower-pitched sounds more in the center of a mix.

In TunePad, you can adjust the pan, gain, and frequency elements of different cells using the mixer interface shown in Figure 7.7. It's also possible to apply these effects in code using Python's **with** construct. Here's an example of a pan effect that shifts the stereo balance of two **playNote** instructions to the far left speaker.

```
1  with pan(-1.0):
2      playNote([ 31, 35, 38 ], beats = 4)
3      playNote([ 31, 35, 38 ], beats = 4)
```

The value of the pan parameter ranges from −1.0 (full left speaker) to 1.0 (full right speaker). A value of 0.0 evenly splits the sound. The **with** keyword in TunePad applies the pan effect to all of the statements indented directly below it.

Figure 7.7 The mixing interface in TunePad allows you to adjust gain, pan, and frequency response for each track in a mix.

7.4.3 Gain

The **gain** of a track is related to its loudness. Gain isn't quite volume, but works as kind of a multiplier to an audio signal's amplitude. When mixing boards were physical pieces of equipment, gain related to the amount of power a signal had at each stage of the signal flow. Now, gain has a similar meaning and can be used to make a track more or less prominent. As an example, a producer may choose to make the bass drum of a dance track more prominent while decreasing the gain of the vocal melody. Gain is commonly measured in decibels. Negative values reduce the loudness of a track, and positive values increase it from its original volume.

7.4.4 Frequency bands

When mixing tracks together, it's often helpful to break the full frequency spectrum into **bands** that correspond to different ranges of frequencies. Each band is meant to capture a particular musical element, although, of course, this varies between genres and specific songs. Producers often split a mix into seven bands: sub-bass, bass, low midrange, midrange, upper midrange, presence, and brilliance. One reason to think in terms of bandwidth is that when sounds have the same bandwidth an acoustic phenomenon called "masking" can occur. Masking is when one sound overpowers another sound such that the sound that is overpowered is not audible.

Band	Frequency range	Description
Sub-bass	20–60 Hz	Adds power and deepness to bass and drums
Bass	60–250 Hz	Captures core fundamentals of the bass and drum sound
Low midrange	250–500 Hz	Captures the overtones of lower instruments, as well as instruments like viola and alto saxophone
Midrange	500–2,000 Hz	Captures melodic instruments such as violin, flute, and the human voice
Upper midrange	2,000–4,000 Hz	Captures overtones of melodic instruments as well as the core of some higher instruments
Presence	4,000–6,000 Hz	Captures the overtones of higher instruments as well as adding precision and clarity to sounds
Brilliance	6,000–20,000 Hz	Captures the upper overtones of all instruments

Humans are most sensitive to frequencies between 1 kHz and 4 kHz. Looking at the frequency values for each band, you might notice that the frequency ranges are not even close to the same size. For example, the sub-bass band covers a range of only 40 Hz (from 20 Hz to 60 Hz), while the presence band covers 2,000 Hz (from 4,000 Hz to 6,000 Hz). The reason is that human perception of pitch isn't linear. When we move up one octave, we are doubling the frequency of a pitch, which means that each consecutive musical octave covers double the frequency range (or bandwidth) of the octave below it. As a result, higher frequency bands naturally cover larger portions of the frequency range and generate more energy.

7.5 Filters and equalization

If two instruments overlap in their natural pitch range, it can be difficult to distinguish one from the other, which can lead to muddiness. The
producer will want to ensure that each musical element is distinct and
audible. Think of a painter recreating an ocean scene. The painter wants
each element of the scene to stand out clearly—perhaps the sky, a boat,
the shore, and the ocean itself. If the ocean, the sky, and the land are all
the same shade of blue, a viewer won't be able to interpret and appreciate
the scene.

The most important tools that a producer has to achieve balance across
the frequency spectrum are **filters and equalizers**. These tools reduce
(**attenuate**) or increase (**boost**) certain frequency ranges in a track to
make them more or less prominent in a mix, and producers will often
"carve out" room in the frequency spectrum for each track. The process of
adjusting the levels of the frequency bands within a signal is called **equalization** (or **EQ**). When you adjust the bass and treble dials of the sound
system in your car, you are equalizing the frequencies just as a producer
might while adjusting the sound of an instrument.

You can think of an audio filter kind of like a filter that you would use
to purify drinking water. A water filter is designed to let small particles
(like water molecules and minerals) pass through while blocking larger
particles (like bacteria). An audio filter achieves a similar effect except
for sound, allowing certain frequencies of an audio signal to pass through
unaffected while blocking or reducing other frequencies. A filter's **response curve** is a graph that shows which frequencies are allowed to
pass through and which are filtered out. There are several types of filters
commonly used in music production including lowpass, highpass, lowshelf, highshelf, bandpass, notch, and peaking. We describe several of
these filters below along with an example of applying these filters with
Python code in TunePad. Most production software (including Tune-
Pad) include built-in equalizer tools that let you combine and precisely
adjust various filter types. Understanding how each filter works will help
you use these tools.

7.5.1 Lowpass filter

A **lowpass filter** allows frequencies below a certain threshold—called the
cutoff frequency—to pass through unaltered. Frequencies above this threshold are reduced (or attenuated). A frequency parameter specifies the location of the cutoff, and a Q parameter determines how sharp or steep this
cutoff is (Figure 7.8).

Lowpass filters might be applied if a track sounds too bright, or to remove some of the higher partials of a bass instrument to make room for
other instruments in a mix, or even to remove some unwanted studio
sound such as buzzing from equipment.

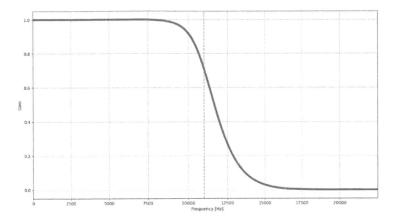

Figure 7.8 Lowpass filter response curve.

In TunePad, you can add a lowpass filter directly in Python code. The example below applies a constant lowpass filter with a cutoff of 100 Hz to reduce higher frequencies in the drums.

```
1  with lowpass(frequency = 100):
2       playNote(0, beats = 1)
3       playNote(2, beats = 1)
4       playNote(0, beats = 1)
5       playNote(2, beats = 1)
```

The **with** keyword starts a special Python structure that applies an effect to all of the statements indented below it. In this case, TunePad's lowpass filter is applied to the four drum sounds.

All of the filters that you can code in Tunepad have the same basic structure. You use the **with** keyword followed by the filter name. Filters have one required parameter and several optional parameters. The only required parameter is frequency, which represents the cutoff frequency for each filter. Filters also have an optional Q parameter, which specifies how sharp or spread out the frequency cutoff is around the target frequency.

Parameter	Description	Required?
Frequency	Cutoff or central frequency specified in Hz	Yes
Q	Typically the sharpness of the cutoff frequency	No
Beats	How long the effect lasts in beats	No
Start	How long to delay in beats before starting the effect	No
Gain	Some filters like peaking, lowshelf, and highshelf use a gain parameter to specify the intensity of the boost or attenuation in decibels	No

7.5.2 Highpass filter

A **highpass filter** is the opposite of a lowpass filter; it passes frequencies *above* the cutoff and reduces frequencies below. As with lowpass filters, the frequency parameter sets the cutoff frequency and the Q parameter specifies the sharpness of the cutoff (Figure 7.9).

A highpass filter might be applied if a track sounds muddy because it has too much bass, or to remove unwanted noise such as a low hum from equipment. Here's a TunePad example that uses a highpass filter to cut out sounds lower than 4,000 Hz (4 kHz) for an instrument playing a melody:

```
1  with highpass(frequency = 4000):
2      playNote(31, beats = 0.5)
3      playNote(35, beats = 0.5)
4      playNote(38, beats = 1)
5      playNote(36, beats = 1)
```

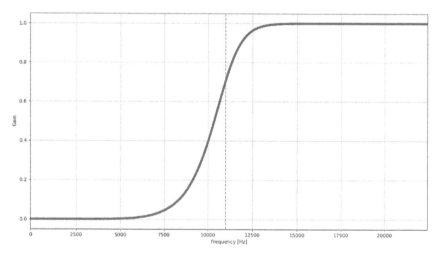

Figure 7.9 Highpass filter response curve.

7.5.3 Bandpass filter

A **bandpass filter** reduces the frequencies above and below a specified band of frequencies; a bandpass is the equivalent of applying both a lowpass and a highpass filter. We specify the middle of the band using the frequency parameter and the width of the band using the Q parameter. The higher the Q, the sharper the cutoff, and the narrower the band of frequencies that can pass through. Bandpass filters allow us to precisely target a track's frequency range. You might use a bandpass filter to bring out the vocals or melody of a song by reducing everything else (Figure 7.10).

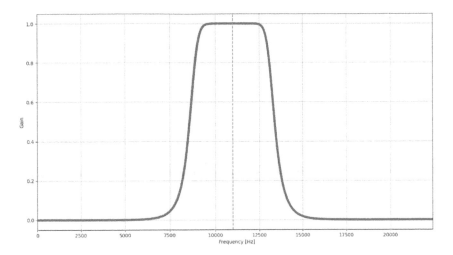

Figure 7.10 Bandpass filter response curve.

In the example below, we will apply a constant bandpass filter with a center frequency of 130 Hz (approximately C3 or MIDI 48) to a short melody to bring the melody out in the overall musical texture.

```
1  with bandpass(frequency = 130, Q = 0.7):
2      playNote(48, beats = 0.5)
3      playNote(52, beats = 0.5)
4      playNote(55, beats = 1)
5      playNote(53, beats = 1)
```

7.5.4 Notch filter

A **notch filter** is the opposite of a bandpass filter. Rather than bringing out a band of frequencies, a notch filter *reduces* the frequency band while all other frequencies pass through freely. Like with the bandpass, the frequency parameter specifies the center of this frequency band and the Q parameter sets the width (Figure 7.11).

In the example below, we will apply a constant notch filter with a center frequency of 440 Hz (approximately A4 or MIDI 69) to a short selection of chords to reduce the prevalence in the overall musical texture.

```
1  with notch(frequency = 440):
2      playNote([69, 72, 76], beats = 4)
3      playNote([69, 72, 76], beats = 4)
```

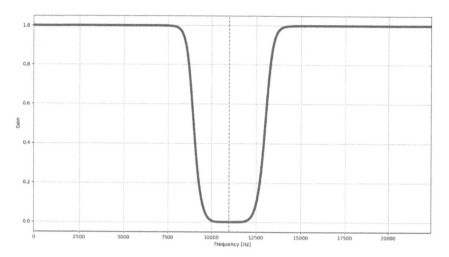

Figure 7.11 Notch filter response curve.

7.5.5 Peaking filter

Peaking filters are frequently used in **parametric equalizers** to boost or attenuate sounds at a target frequency. Parametric equalizers are a type of equalizer that offer precise control of the center frequencies and Q (how spread out or tight the filter is around the center frequency). Using these filters, there's a third parameter called gain that controls how much the

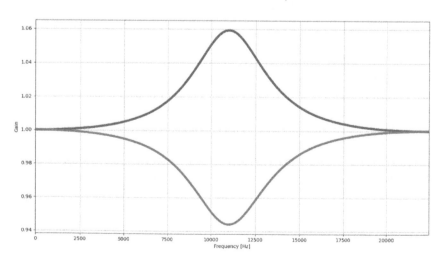

Figure 7.12 Peaking filter response curve.

signal is boosted or attenuated. Gain is measured in decibels. A positive gain will boost the frequencies targeted by the filter, and a negative gain will attenuate them (Figure 7.12).

7.5.6 Lowshelf and highshelf filters

The lowshelf and highshelf filters boost or attenuate sounds beyond the target frequency. They are called shelves due to the plateau shape of their response curves. As with peaking filters, the frequency parameter specifies the cutoff, and the gain parameter specifies how much boost or attenuation to give to frequencies beyond the target. Negative gain values attenuate and positive gain values boost (Figure 7.13).

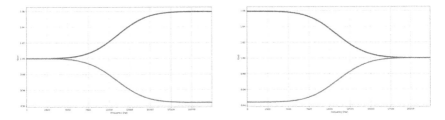

Figure 7.13 Low shelf and high shelf response curves.

7.6 Mastering

After the individual tracks have been adjusted in relation to one another, **mastering** is the process of taking this final mix and polishing it by adjusting global parameters such as dynamic range and frequency. In the early days, there were mastering engineers who specialized in the process of mastering final mixes. In fact, there were studios dedicated to mastering, so you can imagine that this last leg of the production process deserves as much attention as the rest. Mastering is particularly important because you want your mix to sound good on as many devices as possible, so there is a delicate process of balancing the elements in the mix to optimize the listening experience across different media. You want your mix to sound as good over speakers as it does over headphones.

Traditionally, mastering is done using tools like equalization, compression, limiting, and stereo enhancement. Recall from Chapter 3 that dynamic range refers to the difference between the quietest and loudest volumes in a selection of audio. This can be adjusted through the use of **dynamic range compression,** or **compressors**. Compressors reduce the highest volumes in a mix and amplify the lowest volumes, which shrinks the overall dynamic range of the audio. This ensures that the listener can hear the full range of volumes clearly. You can think of this like an action movie where a character might whisper a secret right before an explosion. Dynamic range compression is one possible tool that could make sure that both of these sounds are clear to the audience by reducing the volume of the explosion and increasing the volume of the whisper.

The producer also considers the frequency domain when creating the final product. Instead of thinking on a track-by-track level as in mixing, the producer can think in terms of different bands of frequencies. By this stage, our different bands of frequencies should already be well balanced, and the goal is to polish the overall mix using EQ and filters.

Lastly, a final version of the track is generated and exported into the final format. A major consideration is where and how the music is going to be distributed. A producer might think about a person watching a music video on a laptop through YouTube, versus someone listening to the radio in a car, versus someone streaming audio online, versus someone with a physical CD or even vinyl recording. Music for streaming and other forms of distribution is almost always **compressed**, meaning that the size of the final audio file is much, much smaller than the original uncompressed audio data. Note that this is not related to dynamic range compression. Compressing audio means that some of the data is discarded to decrease the amount of information that has to be transmitted over the internet to avoid buffering delays or to store more songs on a CD. There are complex computer algorithms that decide what data is discarded so that listeners won't even notice a reduction in quality. Examples of file formats that use this form of compression include .MP3 and .AAC files.

Some audio formats forgo this compression in favor of increased sound quality and fidelity. These files contain the raw audio data. These audio files are generally larger, taking up more file space. Examples of this include the .WAV and .AIFF file formats.

Mixing and mastering can be a tedious process requiring attention to detail and a keen, well-trained ear. Becoming an accomplished professional can take many years of experience, and we've introduced just a few of the parameters and tools at your disposal. Don't stress over this, especially at first! Mixing and mastering are two of the most difficult concepts in producing music, but having a grasp of them can greatly elevate the music that you create. The best way to gain this familiarity is through experimentation and thoughtful listening. Listening critically to music that has been professionally produced will help develop your ear and unlock a whole new world of possibilities to your music.

7.7 Dynamic effects in TunePad

Many of the TunePad effects described above can also be applied dynamically to create a wide variety of sounds. The basic idea is that instead of passing one constant value for a parameter, we instead pass a list of numbers that describes how that parameter will change over time. The duration of the dynamic effect is specified using a **beats** parameter. You can try these dynamic effects and filters for yourself at

https://tunepad.com/examples/effects.

Here's an example that uses a lowpass filter to create a wha-wha effect at the beginning of a piano note. The cutoff frequency of the filter moves rapidly back and forth between 200 Hz and 800 Hz over the duration of 1 beat. The note itself plays for three beats, so only the first beat of the note has the effect applied:

```
1  # creates a wha-wha effect by quickly changing
2  # the cutoff of a lowpass filter between 200 and 800hz
3  with lowpass(frequency = [200, 800, 200, 800, 200, 800], beats=1):
4      playNote(47, beats=3)
```

The other effects like pan and gain work the same way. You can create dynamic changes by passing in a list of values and **beats** parameters. For example, this code gradually sweeps a lowpass filter from 20 Hz to 750 Hz and back again. At the same time, it moves the sound across the stereo field from the left to the right and back again. To do this, it nests the pan effect inside of the lowpass filter effect.

```
1  with lowpass(frequency = [20, 750, 20], beats = 40):
2      with pan(value = [-1, 1, -1], beats = 40):
3          playNote(16, beats=40)
```

The interlude following this chapter shows how to add other dynamic effects to TunePad projects using Python code.

Notes

1 This use of the word "track" has a different meaning than a track on an album.
2 Recordings with only one channel of audio data are called monophonic, or **mono**, recordings.

Interlude 7
CREATIVE EFFECTS

In this interlude we're going to work with audio effects—such as filtering and gain—as creative compositional tools. In Chapter 7, we saw how static effects like filters and gain can be used in the mixing process. These same effects can also be used as compositional tools to help craft a cohesive soundscape. They can build tension, add contrast, and drive a song forward. You can follow along online with this TunePad project: https://tunepad.com/interlude/effects.

OPTION 1: Fades

Fades are one of the most common creative effects. You'll often hear fades in or out at the beginnings or ends of songs. These fades are essentially gradually moving from one volume to another: for a fade-in, low to high; for a fade-out, high to low.

We can easily add fades to our music in TunePad using the gain effect and a list. If the gain class is passed a list, it moves evenly from each value to the next, over the whole duration of beats. For a fade-in, we want to move from silent to full volume. Say we have a function called **phrase** that plays eight beats of a melody. We can use the gain class with a list as the values input:

```
1   with gain([0.0, 1.0], beats = 8):
2       phrase()
```

DOI: 10.4324/9781003033240-14

Or, if we want to fade out our phrase:

```
1    with gain([1.0, 0.0], beats = 8):
2        phrase()
```

What if we want to control the speed of our fade? Because gain moves smoothly from each value, we can shape our fade by adding more intermediate values between 0.0 and 1.0. You can think of this as specifying more points along the curve. If we want our fade-in to be quieter for longer, we could add additional values closer to zero:

```
1    with gain([0.0, 0.05, 0.1, 0.15, 1.0], beats = 8):
2        phrase()
```

We can see the difference graphically below (Figure 7.14).

With method 1, the ramp-up in volume is consistent and gradual over the entire duration. But with method 2, the ramp-up is slow until the seventh beat. These last two beats have the greatest increase in gain, increasing from 0.15 to 1.0.

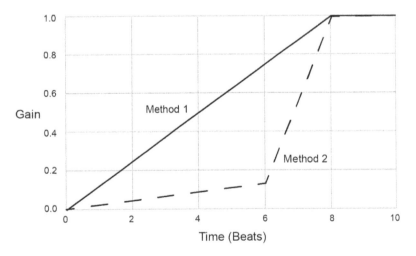

Figure 7.14 Graph of two methods for fading audio in.

OPTION 2: Filter sweeps

Filter sweeps are a great way to build tension in a song. The principle behind a filter sweep is to apply a filter to a section of a song—which blocks some of the frequencies—and then gradually remove it, revealing the full spectrum of the audio. This can use either highpass or lowpass filters, each imparting a different sound to our track. A lowpass filter will start with just the low frequencies and will gradually reveal the upper overtones while a highpass filter will do the opposite.

In TunePad, this is going to look similar to our fades. We'll pass our filter a list of two or more numbers. However, these numbers will represent frequencies, so we need to decide which values to use. Like with our fades, we want a higher value and a lower value. The exact values depend largely on the specific bandwidth of our instrument, the desired speed of our sweep, and personal taste. For our highpass sweep, we want our initial value to block most of the frequencies, and our final value to include most of the spectrum of the audio. Our higher value can be a frequency near the upper range of human hearing; for our lower value, we could choose a frequency closer to the bottom end of human hearing. Let's look now at the code for a basic highpass filter sweep using our phrase function and the values 22,000 Hz and 22 Hz:

```
1   with highpass([22000, 22], beats = 8):
2       phrase()
```

For the lowpass filter sweep, we also want to initially block most of the component frequencies and gradually reveal the upper frequencies of the audio. Due to how lowpass filters work, the lower number should be first. We can use the initial value of 22 Hz to block most of the frequencies and end at 4,000 Hz, revealing most of our spectrum. Let's look now at the code for a basic lowpass filter sweep, again using our phrase function with our chosen values:

```
1   with lowpass([22, 4000], beats = 8):
2       phrase()
```

If we want to have greater control of the shape of our sweep, we can use the same principle behind our gain. Adding more values allows us to better sculpt our resulting sound. If we want our filter to evolve slowly at first, we can add additional values near our starting frequency. Here's an example of this with our highpass sweep:

```
1   with highpass([22000, 18000, 14000, 8000, 22], beats = 8):
2       phrase()
```

Most of the audio spectrum is revealed between the 8,000 Hz and 22 Hz values, which occurs on the last two beats. Try experimenting with different values.

OPTION 3: Reverb

One of the most important effects that producers use is reverb. When sound waves move through a physical space, some of those waves bounce back to the listener. The waves that bounce back are heard as softer. Think of how sound reverberates through a concert hall, or even your bathroom. We can recreate this reverberation through applying reverb to our track.

There are different mathematical strategies for applying reverb to audio. TunePad uses something called **convolution reverb**, which is one of the most common varieties. This takes an audio sample called an **Impulse Response** from a real-world space that represents how different frequencies resonate through the physical space and essentially maps this Impulse Response over our selection of audio.

In TunePad, we specify our impulse response by choosing from a selection of preset values:

- Hall
- Gallery
- Museum
- Library
- Theatre
- Underpass
- Space Echo 2

This is the first argument. Capitalization, spacing, and punctuation are ignored. Like the other effect classes, we can also optionally specify the amount of beats the effect should last and a delayed start parameter. This effect also uses a parameter called **wet**. This specifies how much reverb is applied. A value of 1.0 represents maximum reverb, and a value of 0.0 represents no reverb. You can also pass a list of numbers to create a change over time. Each number will be evenly distributed over the duration of the effect specified by the beats parameter. In the examples below, we are gradually applying the "Underpass" reverb, which is the most reverberant, to our phrase function:

```
1  with reverb(impulse = "Underpass", wet = [0.0, 1.0], beats = 8):
2      phrase()
```

Experiment with different impulse responses!

8 Note-based production effects

This chapter covers a variety of production effects that can add sophistication and depth to your sound. All of the effects are variations on the same theme—instead of playing one note, we'll play a series of notes, each offset slightly in time or pitch. From this basic technique, we'll work through a wide variety of effects including echoes, arpeggiation, chorus sounds, a swung beat, and a phaser. To create these effects, we'll define some of our own functions that make use of TunePad's **rewind** and **fastForward** features. This will also give us a good chance to review loops, variables, and parameters from earlier chapters. Once you master these basic techniques, it will open a range of audio effects that you can expand and customize to define your own unique sound. We'll also cover how to combine these techniques with other effects and filters to add even more flexibility.

8.1 Out-of-tune piano effect

Let's start with one of the more straightforward effects, an out-of-tune piano. As always, you can try this example by visiting
https://tunepad.com/examples/out-of-tune.

Recall from Chapter 3 that the space between separate notes on a 12-tone chromatic scale can be subdivided into even units called **cents**—just imagine 100 individual smaller notes between each adjacent key on the piano keyboard. On an instrument like a violin or trombone, you can

DOI: 10.4324/9781003033240-15

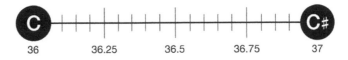

Figure 8.1 Intermediate pitches between C and C♯.

play notes that are slightly out of tune or that glide between one note and another. We can do something similar in TunePad by using decimal numbers instead of integer values when we call the **playNote** function:

```
playNote(36.5) # plays an out-of-tune C
```

For this line of code, a value of 36.5 sits exactly halfway between a C (MIDI value 36) and a C# (MIDI value 37). In other words, it's a pure C *detuned* by 50 cents (Figure 8.1).

There are plenty of artistic reasons to create sounds that are out of tune—like if we wanted to create an eerie melody for a horror movie. To get this kind of effect, we could just randomly **detune** some of the notes in our melody by small amounts, and we'd get something that sounded off key. But a more interesting approach that gives us additional texture (and eerie dissonant overtones) would be to play several notes at the same time, each slightly detuned from one another. This actually approximates what a real out-of-tune piano sounds like. Notes on a piano are produced when a hammer mechanism inside the instrument hits multiple individual strings at the same time (three strings for most notes). When those individual strings are off from one another, you hear an entirely different sound than you would get from just one string being out of tune. Honky tonk pianos are intentionally tuned so that the three strings for each note are detuned slightly from one another to get warped harmonics. Here's a simple Python function that approximates that dissonant sound in TunePad:

```
1  def errieNote(note, beats = 1, velocity = 100):
2      volume = velocity / 3.0
3      for i in range(3):
4          offset = random() −0.5
5          playNote(note + offset, beats = beats, velocity = volume)
6          rewind(beats)
7      fastForward(beats)
```

Let's walk through this function one line at a time. If this example makes sense, all of the other functions that we create later in this chapter should be easier to understand. On line 1, we use the **def** keyword to define our own Python function.

```
1 def errieNote(note, beats = 1, velocity = 100):
```

Flip back to Chapter 4 for an in-depth description of function definitions, but the main thing to remember is that creating our own functions lets us build up our own musical toolbox to help create more complex compositions. In this case we're adding **eerieNote** as our newest tool. We're also defining three *parameters* for this function called **note**, **beats**, and **velocity**, each of which is listed inside the parentheses. We use **note** for the pitch value we want to play; **beats** for the duration of the note; and **velocity** to approximate the overall volume of the sound. You might notice that these parameters are exactly the same as the **playNote** function. That's intentional because it will make it easy to swap out the **playNote** function for our new **eerieNote** function in other parts of our project. One other thing to notice here is that **beats** and **velocity** are examples of what's called an *optional* parameter. That means we've defined a default value that Python will use if we don't otherwise specify something. The default value for **beats** is 1, and the default value for **velocity** is 100. So, for example, each of these lines of code will do the same thing, and we can use them interchangeably:

```
eerieNote(36)
eerieNote(36, beats = 1)
eerieNote(36, beats = 1, velocity = 100)
eerieNote(36, 1, 100)
eerieNote(36)
```

In all of these cases, **beats** and **velocity** are the same as their default values. We can call the function with other values too:

```
eerieNote(40, beats = 0.5, velocity = 50)
eerieNote(40, beats = 1.5)
eerieNote(40, velocity = 120)
eerieNote(40, 2.0, 50)
```

Let's go back to the **eerieNote** function. On line 2 we define a *variable* called **volume** that will help us adjust the loudness of our individual notes.

```
2 volume = velocity / 3.0
```

We're doing this because playing all the notes at full volume would end up being much louder than the sound of a single note. We set its value to the **velocity** parameter divided by 3.0 because we're going to end up playing three notes instead of one, so we want each individual note to make up about a third of the overall volume.

On line 3, we set up our **for** loop to play three notes. You can also experiment with loops that repeat different numbers of times. Remember

that **range** is just a Python function that generates a sequence of numbers
[0, 1, 2] that the variable **i** will walk through, one number at a time.

```
3 for i in range(0, 3):
```

On line 4, we use Python's **random** function to generate a random decimal number somewhere between zero and one. Subtracting 0.5 will then
give us a number in the range of negative 0.5 to positive 0.5. We'll save
the result in a variable called **offset** that we'll use on the next line of the
code to shift our note's pitch.

```
4 offset = random() - 0.5
```

Line 5 then plays the note with our random pitch offset added in (between
−0.5 and 0.5) to make it sound out of tune. Notice that we use our **beats**
and **volume** variables to control the duration and volume of the note that
gets played, just as we might if we were calling **playNote**.

```
5 playNote(note + offset, beats = beats, velocity = volume)
```

The last two lines of the function make it so that all of the notes get played
at the same time. On line 6, we're going to use a TunePad command called
rewind to move the playhead back to where we were before we played
the note. Remember that **playNote** automatically advances the playhead
forward by the given number of beats, so we need to rewind to get us back
where we started. The effect of calling rewind is instantaneous; all it does
is reposition the playhead.

```
6 rewind(beats)
```

Calling **rewind** is important because we want all three notes to play at
exactly the same time to give us the right effect. Later in the chapter, we'll
experiment with playing notes that don't all get triggered at the same time.

Lines 4, 5, and 6 are all part of the **for** loop—they all get repeated three
times in a row because they're indented below the loop statement on line
3. We finish the function outside of the loop with line 7 that calls another
TunePad command called **fastForward**.

```
7 fastForward(beats)
```

As you might guess, this does the opposite of **rewind**—instead of moving the playhead backward it moves it forward. We call this last **fastForward** to make **eerieNote** behave exactly as if we had played a single note
using our standard **playNote** function. The playhead will be moved by
the value of **beats**. This combination of using **rewind** inside of a loop

with a **fastForward** at the end of the loop will be the standard template for all of the remaining effects that we'll cover in the chapter. We'll mix things up, but the basic ideas will be the same.

To put this all together, here's a childhood favorite that merrily reminisces about the black plague. Sticking with the horror movie theme, we're going to make this extra menacing by setting the tempo way down to 60 BPM and playing the whole thing an octave lower than usual. The snippet listed below defines the entire song (including pitches, durations, and lyrics) in a Python list starting on line 1. Each note is represented by a Python structure called a *tuple* that we haven't used in this book until now. Tuples are written as values separated by commas—just like lists except that you enclose the values inside of parentheses instead of square brackets. We can use tuples exactly as we would a list except that the values inside a tuple can't be changed. Saying the same thing in Python lingo, tuples are *immutable* **objects**. Lines 18 and 19 step through the notes in the song, tuple by tuple, calling **eerieNote** for each one. Saying **note[0]** gets the pitch of the note and **note[1]** gets the duration. You could also include this line inside the loop if you wanted to print the lyrics.

```
print(note[3])
```

```
1   song = [
2       (48, 1, "Ring"), (48, 0.5, "a-"), (45, 1, "round"), (50, 0.5, "the"),
3       (48, 1.5, "ro-"), (45, 1, "sie, "), (47, 0.5, "a"), (48, 1, "pock-"),
4       (48, 0.5, "et"), (45, 1, "full"), (50, 0.5, "of"), (48, 1.5, "po-"),
5       (45, 1.5, "sies!"), (48, 1.5, "Ash-"), (45, 1.5, "es!"), (48, 1.5, "Ash-"),
6       (45, 1, "es!"), (45, 0.5, "We"), (48, 1.5, "all"), (48, 1.5, "fall"),
7       (41, 2.5, "down.")
8   ]
9
10
11  def eerieNote(note, beats=1, velocity=100):
12      volume = velocity
13      for i in range(0, 3):
14          offset = (random()-0.5) # random detune amount
15          playNote(note + offset, beats, velocity=volume)
16          rewind(beats)
17      fastForward(beats)
18
19
20  for note in song:
21      eerieNote(note[0] -12, beats=note[1])
```

Here are the note values and associated TunePad commands for the entire song:

Note	Beats	Lyrics	Code	Note	Beats	Lyrics	Code
C 48	1.0	Ring	playNote(48, beats=1)	C 48	1.5	po -	playNote(48, beats = 1.5)
C 48	0.5	a -	playNote(48, beats = 0.5)	A 45	1.5	sies,	playNote(45, beats = 1.5)
A 45	1.0	round	playNote(45, beats = 1)	C 48	1.5	Ash -	playNote(48, beats = 1.5)
D 50	0.5	The	playNote(50, beats = 0.5)	A 45	1.5	es!	playNote(45, beats = 1.5)
C 48	1.5 ♭.	ro -	playNote(48, beats = 1.5)	C 48	1.5	Ash -	playNote(48, beats = 1.5)
A 45	1.0	sie,	playNote(45, beats = 1)	A 45	1.0	es!	playNote(45, beats = 1)
B 47	0.5	A	playNote(47, beats = 0.5)	A 45	0.5	We	playNote(45, beats = 0.5)
C 48	1.0	pock -	playNote(48, beats=1)	C 48	1.0	all	playNote(48, beats = 1.5)
C 48	0.5	et	playNote(48, beats = 0.5)	C 48	1.0	fall	playNote(48, beats = 1.5)
A 45	1.0	full	playNote(45, beats = 1)	F 41	2.5	down.	playNote(41, beats = 2.5)
D 50	0.5	of	playNote(50, beats = 0.5)				

8.2 Phaser effect

If you play the song in the previous example, you might hear some interesting and unexpected effects, especially on the lower notes that come about from playing several notes together with very similar pitches. It turns out that this approximates a common musical effect called a *phaser* or *phase shifters*. Real phaser effects are produced by playing multiple versions of the same sound together at the same time, but changing the frequency profile of each individual sound to get gaps or dips in the spectrum.

A very simple variation of the **eerieNote** function replaces the use of the **random** function and instead just increments the pitch **offset** variable by a fixed amount. This change gives us a cool approximation of a phaser effect that's easy to manipulate by changing the number of simultaneous notes that get played or by changing the pitch offset. Try this effect with some of the built-in TunePad instruments, or use the Sampler instrument to record and playback your own voice with the **phaserNote** function.

```
1    def phaserNote(note, beats = 1, velocity = 100):
2        note_count = 5
3        volume = velocity / note_count
4        offset = 0.0
5        for i in range(0, note_count):
6            playNote(note + offset, beats = beats, velocity =
             volume)
7            offset += 0.15
8            rewind(beats)
9        fastForward(beats)
```

You can try this example online by going to https://tunepad.com/examples/phaser.

8.3 Echo effect

For the next set of examples, we're going to move from pitch-based effects to time-based effects where we spread multiple notes out over time. The simplest version of this is an **echo effect** that plays an initial sound at full volume followed by a rapid succession of softer notes that fall off into silence. This is an extremely common technique used in a wide variety of digitally produced music. We can also throw in pitch manipulations and reverb effects to add another layer of complexity, but let's start with the foundational repeated sound. As before, we're going to define our own function that looks similar to the basic **playNote** function. You can follow along in TunePad by visiting https://tunepad.com/examples/echo.

```
1    def echoNote(note, beats = 1, delay = 0.125):
2        volume = 100        # start at full volume
3        offset = 0          # keep track of the delays
4        while volume > 1:    # loop until silence
5            playNote(note, beats = beats, velocity = volume)
6            rewind(beats - delay)
7            volume *= 0.5    # reduce volume by 50%
8            offset += delay # keep track of the delays
9        rewind(offset)        # rewind by the accumulated delay amount
10       fastForward(beats)    # move the playhead to the end of the note
```

Line 1 looks similar to the previous two examples except that we've added another optional parameter called **delay** that specifies how spread out each echoed note is in time. By default we've set this to

0.125 beats, but by making this an optional parameter, we've given the composer the ability to change this delay time on the fly if they want to.

Line 2 should also look familiar except this time we're going to start with the first note at full volume and then rapidly decay the volume for each successive note. Then on line 3, we define a variable called **offset** that keeps track of the total amount of delay accumulated so far. This is a bookkeeping variable that's important for us to leave the playhead in the correct location after our function completes. We'll use this on line 9.

On line 4, we set up a new kind of Python loop called a **while** loop. Until now, we've always used **for** loops to iterate through a list or repeat something for a *fixed* number of times. With a **while** loop on the other hand, you'll repeat something over and over again *until* a certain condition is met—in this case the loop will repeat until the volume is less than or equal to 1.

```
4 while volume > 1:
```

The basic idea is to repeat the sound until it's too quiet to hear. The trick to making this work is to decrease the value of **volume** inside the while loop. Otherwise, the loop would keep repeating over and over again forever (an infinite loop). We do this on line 7 where we multiply volume by 0.5 (50% of its previous value). We could try other values here or even make this a parameter of the function instead of a hard-coded value. If you want something to echo out longer, you could set this up to 0.6 or even 0.7. Again, be careful here because if this value is 1.0 or higher the note will echo forever and TunePad will complain that you've created an infinite loop! Higher values can also cause the note to echo out too long and run into other notes, creating unwanted interference.

Line 6 is a little tricky. Instead of rewinding all the way back to the beginning of the note as we did in the previous example, we're going to rewind by a smaller amount so that successive notes are spread out in time.

```
6 rewind(beats - delay)
```

This total amount is equal to the **delay** parameter that gets passed into the function. The diagram below shows what this looks like over four iterations of the **while** loop. You can also see the value of the **volume** as it decreases on each successive pass.

The last two lines of the function are responsible for fixing up the playhead position so that it's exactly at the end of the first note we played, which is why we kept track of the number of the total accumulated delay.

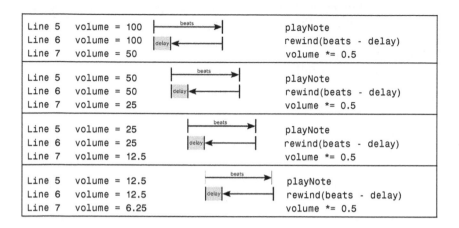

8.4 Chorus effect

A closely related effect to echo is to randomly scatter notes around a central point in time. Our strategy for this effect should look familiar by now. The only tricky part is that we have to generate a random time offset plus or minus a 32nd note that we use to fast forward. Because this number can be positive, negative, or zero, the playhead will move forward, backward, or stay in place when we call **fastForward**. Another way to think of it is that **fastForward** with a negative number is the same thing as calling **rewind** with a positive number. This function takes an optional parameter called **count** that says how many scattered notes we should play.

```
1   def chorusNote(note, beats = 1, count = 4):
2       volume = 40
3       spread = 0.125
4       for i in range(0, count):
5           offset = (random()-0.5) * spread
6           fastForward(offset)
7           playNote(note, beats - offset, volume)
8           rewind(beats)
9       fastForward(beats)
```

Then, when we rewind on line 8, we're going to back up to where we started. Together this has the effect of scattering several notes randomly around the time the note was to be played. You can also try making the spread an additional optional parameter that you pass into the function. Here's a simple stomp/clap pattern that uses **chorusNote**:

```
stomp = [ 0, 1 ]
clap = 22
chorusNote(stomp)
chorusNote(stomp)
chorusNote(clap)
rest(1)
```

This is a great effect to combine with a reverb effect to make it sound like you have a crowd in a large room. Another common technique is to detune the pitch of each note slightly from one another, which was a common technique for chorus effects applied to electric guitars in the 1980s. You can try this function by going to https://tunepad.com/examples/chorus.

8.5 Arpeggiation effect

Another extremely common effect called arpeggiation is used to transform single sustained notes into rhythmic patterns that rapidly climb up and down a chord structure. We saw this in Interlude 4 as a method of playing chords. Arpeggios are commonly used across a wide variety of musical genres from classical to hip hop to dance music, and most digital audio workstations (DAWs) provide built-in arpeggiators with a variety of options to change the speed, direction, and note pattern. Following the pattern we saw with previous functions, we can build an arpeggiator with TunePad by creating an **ARPNote** function with parameters for the note pattern and speed. This effect combines both pitch-based and timing-based variations of notes.

```
1   def ARPNote(note, beats = 1, pattern = [0,4,7], speed = 0.125):
2       offset = 0
3       while offset < beats:
4           for step in pattern:
5               playNote(note + step, speed)
6               offset += speed
7       rewind(offset)
8       fastForward(beats)
```

Before we put this together in a complete example, let's look at the **pattern** parameter. This parameter is a list that describes a pitch offset from the first note (given by the **note** parameter). The function will step through this list, and each element will be added to **note** to calculate which note to play. The **pattern** parameter can be as simple or complex as you want, but a good starting point are some of the chords we introduced in Chapter 3. Here are some common arpeggiation patterns:

```
Major Triad              [ 0, 4, 7 ]
Minor Triad              [ 0, 3, 7 ]
Major Triad Up/Down      [ 0, 4, 7, 12, 7, 4 ]
Minor Triad Up/Down      [ 0, 3, 7, 12, 7, 3 ]
```

This function is a great example of using *nested* loops in Python, which just means that one loop is embedded inside of another. The outside loop gets set up on line 3 and repeats until the variable **offset** is greater than the duration of the note. The inner loop is set up on line 4 and steps through each value in the **pattern** parameter. What all this means is that the inner for loop will get run at least once and possibly many times—as long as it takes to completely fill the duration of the note provided by the **beats** parameter. Line 5 plays the notes in the arpeggio by adding **step** to the base note such that it moves up or down the pattern, and we set the duration of the note to speed. Then on line 6, we increment the **offset** bookkeeping variable by the duration of an arpeggiation step. The last two lines of the function clean up the timing to make sure the playhead advances by the correct amount.

The **speed** parameter is a matter of taste and musical convention, but it's common to use 16th notes (0.25 beats) or 32nd notes (0.125 beats) for this value. One thing you might notice is that this function will repeat the arpeggiation chord pattern for as long as it takes to fill out the entire note duration, but it might also play out beyond the end of the note. This extra holdover could be undesirable depending on the music we're trying to make. To fix this, we can insert one additional line right after line 6 inside the for loop.

```
if offset >= beats: break
```

This has the effect of exiting out of the loop as soon as we hit the desired length, possibly short circuiting the ARP pattern. Before we look at an example of the **ARPNote** function in action, let's add one more feature. It sounds a little heavy to have every note of the arpeggio hit with the same velocity. To add a more interesting rhythm and texture, we could instead emphasize the first note of the arpeggio and then drop the volume down for the remaining notes. Here's a variation of the **ARPNote** function that creates this effect with something like the technique we used in the **echoNote** example.

```
1    def ARPNote(note, beats = 1, velocity = 100, pattern = [0,4,7], speed = 0.125):
2        offset = 0
3        while offset < beats:
4            volume = velocity            # reset volume every iteration
5            for step in pattern:
6                playNote(note + step, speed, volume)
7                volume = velocity * 0.25 # reduce volume after first note
8                offset += speed
9                if offset >= beats: break
10       rewind(offset)
11       fastForward(beats)
```

What's happening is that we create a **volume** variable that gets reset every iteration of the while loop to the full value of the **velocity** parameter. Then, after we play the first note of the arpeggio inside the for loop, we reduce the **volume** to a quarter of the **velocity** for the remaining notes in the arpeggio. When the arpeggio is finished, **volume** is restored to its original value for the next arpeggio.

This project puts everything together with an example from The Police. Try this code at code at https://tunepad.com/examples/arp.

When you try this online, you might notice that we've added one more trick to the **ARPNote** function on line 5. This uses the **sustain** parameter of **playNote** to have the first note of the arpeggio ring out for the entire duration of the set, but at a reduced volume. This gives the arpeggio a nice resonant quality.

8.6 Swing effect

The last effect in this chapter creates a swing beat from a standard straight rhythm. Essentially this means adding a bounce or swing to a rhythm's timing so that we move from mechanical precision to more of a human feel. Most DAWs provide a variety of options for swinging a beat. Typically this involves altering a beat at either the 8th or 16th note level so that odd-numbered notes are stretched out slightly in time, while even-numbered notes are compressed by the same amount. Most DAWs let you select the ratio between even- and odd-numbered notes by either offering a selection of fixed choices or a continuous dial. Creating something similar in TunePad turns out to be fairly straightforward. The hardest part is figuring out whether we're on an even or an odd beat. To do this, we can use the built-in TunePad function called **getPlayhead()** that returns the current position of the playhead in beats (quarter notes). To figure out which eighth note we're on, we can divide by an eighth note duration (0.5 on line 3 below). We use Python's **round** function to make sure this is an integer value. The next step is to ask whether we're on an even or an odd eighth note. We do that in line 4 using the modulus operator (%). You can think of this as returning the remainder of a division operation. In other words, divide the step variable by 2. If the remainder is 1, it's an even eighth note, and we need to push the note forward a little by the swing factor that we defined on line 2. Otherwise, it's an odd eighth note, and we leave it in place. Together this has the same effect as stretching out the duration of the odd-numbered notes.

```
1   def swingNote(note, beats=1, velocity=100):
2       swing = 0.25
3       step = round(getPlayhead() / 0.5)
4       if step % 2 == 1:
5           fastForward(beats * swing)
6           playNote(note, beats, velocity)
7           rewind(beats * swing)
8       else:
9           playNote(note, beats, velocity)
```

This example codes a simple rock beat that uses **swingNote**. Try this online at https://tunepad.com/examples/swing. Try adjusting the value of the swing factor on line 2 to get a feel for how it adds bounce to the rhythm. Set swing down to 0.0 to get a straight beat.

Interlude 8
HOW TO MAKE A DRUM FILL

In this interlude, we explore four kinds of drum fills. A **drum fill** is a short phrase dropped into the main groove of a drum track, usually every 8 or 16 bars. Fills add variety and support the transition between sections of a song (verse to chorus, for example). You can follow along online with this TunePad project: https://tunepad.com/interlude/drum-fills.

STEP 1: The groove

Let's start by creating a four-beat **groove**. For this tutorial, we'll use a sparse rock-style drum pattern. We're keeping it simple because we're going to add decoration with the various fills. The code below wraps the drum pattern inside a function definition so that it's easy to reuse in the examples below. Define the **groove** function inside of a **Code** cell in TunePad so that we can import it into other cells.

```
def groove():
    playNote(0, beats = 0.5)
    playNote(4, beats = 0.5)
    playNote(2, beats = 0.5)
    playNote(4, beats = 0.5)
    playNote(0, beats = 0.5)
    playNote([0, 4], beats = 0.5) # double kick with hat
    playNote(2, beats = 0.5)
    playNote(4, beats = 0.5)
```

DOI: 10.4324/9781003033240-16

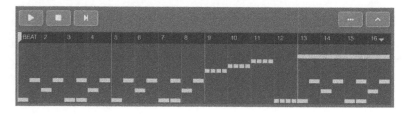

Figure 8.2 Code cell in TunePad showing import statement.

Figure 8.3 Drum fill pattern A.

Set the title of your code cell to "groove" and then note the Python import statement that it generates for us directly below the title (Figure 8.2).

Pattern A: Tom runs

Let's start with the easiest drum fill. This fill takes up one measure (four beats) and consists of 16th note runs on the high tom, mid tom, low tom, and kick drum in order (Figure 8.3).

We can code this with simple for loops to play each of the drum sounds:

```
1    from groove import * # import our groove function
2
3    def fillA():
4        for i in range(4):
5            playNote(6, beats = 0.25) # high tom
6
7        for i in range(4):
8            playNote(7, beats = 0.25) # mid tom
9
10       for i in range(4):
11           playNote(8, beats = 0.25) # low tom
12
13       for i in range(4):
14           playNote(0, beats = 0.25) # kick drum
15
16       playNote(9, beats = 0, sustain = 4) # crash cymbal
```

The fill finishes with a crash cymbal hit on **line 16**. We use the **sustain** parameter to let the crash cymbal ring out and overlap with the next drum measure. Here's the code to play the fill:

```
# let's try it!
groove()
groove()
fillA()
groove()
```

Pattern B: Triplets

Our next pattern plays triplet notes that subdivide each beat into three equal parts. The pattern repeats four times in a row with HIGH TOM – LOW TOM – KICK DRUM combos (Figure 8.4).

We can code this with just one for loop where we set the beat duration to 1/3.0.

```
1  from groove import *
2
3  def fillB():
4      for i in range(4):
5          playNote(6, beats = 1/3.0)
6          playNote(8, beats = 1/3.0)
7          playNote(0, beats = 1/3.0)
8      playNote(9, beats = 0, sustain = 4)
```

Use this code to play **fillB** with the groove.

```
groove()
groove()
fillB()
groove()
```

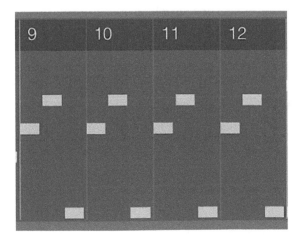

Figure 8.4 Drum fill pattern B.

Pattern C: Random 16ths

The next pattern sprinkles random 16th note hits in to decorate an otherwise boring pattern.

To do this we're going to use Python's **choice** function which is part of the built-in **random** module. This function works by picking an element from a list at random. You can think of it as drawing a random card from a deck. To use **choice**, we have to add a line at the top of our code to import the function:

```
from random import choice
```

The next secret ingredient is to define a list of possible drum sounds that will get selected at random. One trick is to pad this list with a few **None** values, which will result in random empty spaces in our pattern. You can experiment with different drum sounds in this list or with using a larger or smaller number of **None**s. You might also notice that we doubled up on the number of ten notes (claps) because we like how they sound.

```
notes = [ 0, 2, 3, 4, 6, 7, 8, 10, 10, 11, None, None, None, None ]
```

Once we've defined our note array, the function is simple to write. We just select and play eight notes from our list at random.

```
1   from random import choice
2   from groove import *
3
4   notes = [ 0, 2, 3, 4, 6, 7, 8, 10, 10, 11, None, None, None, None ]
5
6   def fillC():
7       for i in range(8):
8           playNote(choice(notes), 0.25)
9       rewind(2)
10
11  # play it with our groove
12  for i in range(10):
13      groove()
14      fillC()
15      groove()
```

There's one trick on line 9 where we rewind the playhead by two beats. We do this so that the drum fill overlaps with the beat instead of pausing it while the fill plays.

Pattern D: Random triplets

Our last pattern is a combination of Pattern B and C. We use triplets but with random notes instead of a fixed pattern. We also use the **choice** function again, but with a smaller set of notes:

```
notes = [ 0, 6, 7, 8, 10 ]
```

Then we can play six triplets (two beats) at random followed by a crash.

```
def fillD():
    notes = [ 0, 6, 7, 8, 10 ]
    for i in range(6):
        playNote(choice(notes), 1/3.0)
    playNote(9, beats = 0, sustain = 4)
```

Here's the code to play this fill with our groove.

```
for i in range(10):
    groove()
    groove()
    fillD()
    groove()
```

9   Song composition and EarSketch

Guest chapter by Lauren McCall

The art of sampling and remixing has become a central practice in modern digital music production. Artists from Dr. Dre to Cardi B have used new technologies to blend short samples from existing music into entirely new compositions. This chapter introduces sampling, remixing, and song composition using a free online platform called EarSketch. This platform is similar to TunePad in that it combines Python programming with music creation. But with EarSketch the focus is more on creating full-length songs by combining and remixing pre-recorded samples instead of composing music from individual notes and percussion sounds. We'll cover the basics of working with EarSketch and show how to structure musical samples into full-length compositions.

9.1 Song structure

There's more to writing music than coming up with a beat, harmony, and melody. Nearly every popular song follows some kind of codified structure. In the days before recorded audio, structure and repetition helped listeners develop relationships with music. Material would get introduced early in a song and then elaborated throughout. The structure of a Blues song is an excellent example, but this is true across almost every genre of popular music from punk to hip-hop to country. Structure and repetition is equally important to the music we listen to today in that it gives us an opportunity to notice contrasting elements, remember a melody, or sing along with the chorus.

DOI: 10.4324/9781003033240-17

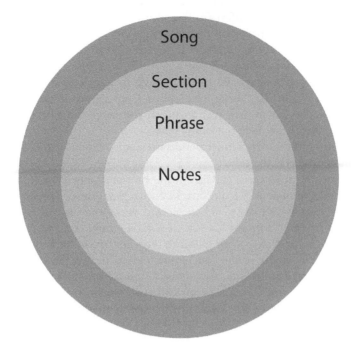

Figure 9.1 Songs are composed of nested and repeating notes, phrases, and sections.

Song structure—sometimes known as **musical form**—describes how musical ideas and material play out over a piece of music. Earlier in the book we introduced the idea of the musical phrase: a self-contained musical thought, like a sentence. We can group these phrases into larger musical sections. These sections can repeat to form even larger structures (Figure 9.1).

The most common way for musicians to classify and discuss different parts of a song is to label them with letters. Every time there's a new section, it gets the next letter in the alphabet. If a section reappears, we reuse the letter we had already applied to that section. If a section reappears but is slightly different—maybe with different lyrics—we can give it a number. This works as a timeline or map of a song. In popular music, **verse-chorus form** is the most common song structure. At its simplest, this form is built on two repeating sections called the **verse** and the **chorus**. There are endless variations on this basic structure, but the form tends to include:

Intro	A1 Verse	B Chorus	A2 Verse	B Chorus	C Bridge	B Chorus	Outro

Some songs will feature an introduction that builds up energy and introduces the musical material. The verse is a repeated section that helps tell the story of a song. Generally, the melody of the verse is the same each time, but the lyrics are different. The chorus, or hook, usually contains an important musical or lyrical motif to the song. This is usually the most memorable part of a song, and the lyrics are almost always the same for each repetition. The bridge provides contrast with the rest of the song, often using something special like a change in harmony or tempo. Not every song has a bridge; some have another repetition of the verse. The final section, the outro, is a way to finish the song. This might include something like a slow fade out of the chorus.

Take a song like Carrie Underwood's hit single, *Before He Cheats* (2005). This follows exactly the structure in the table above. There's a short intro followed by the first verse ("Right now, he's probably slow dancin'..."), followed by the chorus ("I dug my key into the side of his pretty little souped-up four-wheel drive..."). A second verse and chorus is then followed by a bridge ("I might have saved a little trouble for the next girl"), one last chorus, and the outro ("Oh, maybe next time he'll think before he cheats"). Like many of the concepts discussed in this book, these rules are more of rough guidelines than hard rules. Being aware of song form can make writing music much easier, and you can think in larger sections. Using variations of common forms gives listeners an entry point as to what to expect from your music.

9.2 Sampling

Sampling involves taking sections from existing pieces of music and repurposing or remixing them in order to make a new creation. A sample might consist of a specific musical element of the original song such as a bassline, drum break, vocals, or even a snippet of speech. Samples often have creative effects applied, such as reverb, chorus, or filters, and they might be looped, pitch shifted, or even reversed. Examples include songs like Kanye West's *Good Morning* (2007), which sampled Elton John's *Someone Saved My Life Tonight* (1975) or *Naughty* By Nature's O.P.P. (1991), which sampled multiple elements from the Jackson 5's *ABC* (1970). Sampling often occurs in dialogue with or homage to the original work. An artist might include a reference to a work that inspired them by incorporating signature chord progressions, instrumentation, or melodies.

The availability of new recording technology in the 20th century gave artists the opportunity to engage with prior generations of music in innovative ways. Although sampling is common to many genres of music, it's one of the defining characteristics of hip-hop music. The roots of hip-hop were in live performance—DJs manipulated existing records on turntables to create completely new soundscapes. Techniques from live performance then carried over into original works driven by advances in recording technology and equipment specifically designed for sampling. Advanced sampling tools are now built into digital audio workstation (DAW) software in which the vast majority of modern music is created.

When combining samples in your composition, be mindful of how the different musical elements work with one another. If your song is in E major and you add a sample that's in D minor, it might clash harmonically. If the sample is in a different tempo, it might not line up rhythmically with the rest of your song. Samples get manipulated in pitch or tempo or a variety of other ways to make them work stylistically with the rest of a song.

COPYRIGHT LAW

Be aware of copyright issues around music that you sample, especially if you're thinking about sharing it online or licensing it to make money. Most of the music on the TunePad and EarSketch websites is licensed so that you can use them however you want, but it is important to get permission or a license when sampling other artists' music.

9.3 Introduction to EarSketch

EarSketch (Figure 9.2) is a free online platform for creating music with code that was developed by researchers at the Georgia Institute of Technology in Atlanta. EarSketch works with a few different programming languages including Python and JavaScript. All of the Python concepts you've been learning about (lists, loops, variables, functions, and parameters) still apply, although the music-making functions are different. For example, EarSketch doesn't have a **playNote** or **rest** function. Instead, it has a **fitMedia** function that places musical samples on the timeline of a DAW.

JavaScript is another important and widely used programming language. It's often called the language of the web because web sites use JavaScript to add logic, user interaction, and dynamic effects. Every time you click on a button on a web page, it's almost certainly running JavaScript code. If you were to look under the hood at TunePad and EarSketch, you would see that they both use JavaScript to generate and play music. If you're interested in trying it out, EarSketch has excellent resources for learning JavaScript.

Once you log into the main EarSketch site at https://earsketch.gatech. edu, you can write programs in the *Code Editor* (Figure 9.2 center). Note

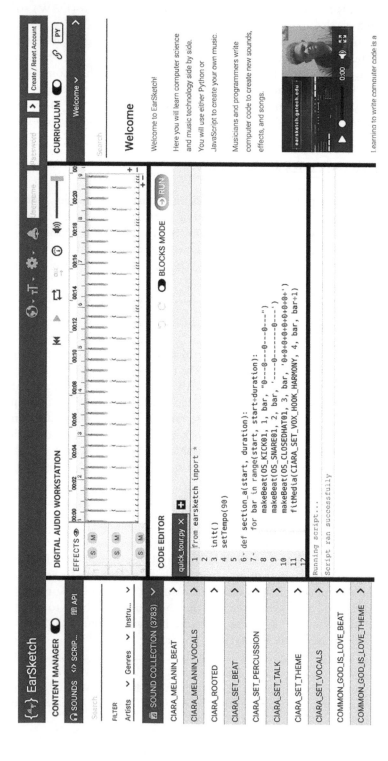

Figure 9.2 The main EarSketch interface features a large library of samples (left), an interactive timeline (middle top), a code editor (center), and extensive documentation and curriculum (right).

that both EarSketch and TunePad sometimes use the term *script* to refer to a program. A script is just another word for a short computer program that performs smaller tasks like putting together a song. Creating a new script in EarSketch generates a small amount of **boilerplate** code to set up your project. In computer programming "boilerplate" means standard code that you use across multiple projects. Here's the EarSketch boilerplate:

```
from earsketch import *

init()
setTempo(120)


finish()
```

The first line imports the core EarSketch module. In TunePad we've used Python's import functionality as well, but you can read more about how it works in the Python appendix at the end of the book. The second line, **init()**, sets up the DAW. Next, **setTempo** specifies the project's tempo in beats per minute (bpm). The code you write for your project should be added in between the **setTempo** and **finish()** function calls.

One of the best parts about EarSketch is its extensive sound library. This library has nearly 4,000 premade audio clips created by producers and musicians that you can use in your projects for free. You can browse through the sample library on the main EarSketch page, filtering by musical genre, artist, and instrument type (Figure 9.2; bottom left). You can add samples to your project with code using their predefined variable names and the **fitMedia** function.

9.3.1 The fitMedia function

The **fitMedia** function positions a sample on a track at the measure of your choice. Figure 9.3 shows a close-up of EarSketch's DAW interface. Individual samples are shown as waveforms laid out on numbered horizontal tracks that give you the ability to organize your composition. For example, you might have separate tracks for your melody, bassline, and harmony. A track can play only one sample at a time with no overlaps. If your code places two audio samples so that they overlap, EarSketch will give you an information message and remove one of the samples from the track.

The **fitMedia** function has the following required parameters:

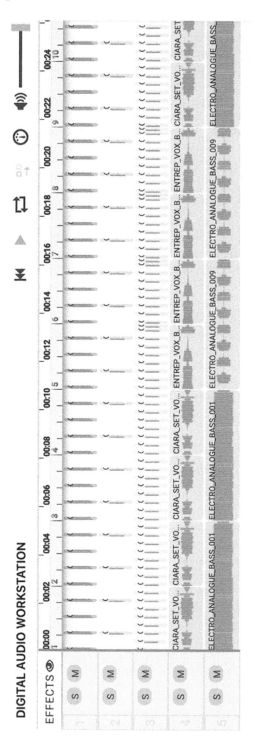

Figure 9.3 Close-up screenshot of EarSketch's DAW.

Parameter	Description
`audio_file_name`	Sample variable name from the EarSketch library -*or-* a string of an audiofile name. If the tempo of the chosen sample doesn't match the tempo of your project, EarSketch will automatically adjust the tempo for you.
`track_number`	Integer value greater than or equal to 1 that specifies which track to put the sample on.
`start_measure`	Decimal value greater than or equal to 1 that specifies which measure to place the sample on. A value of 1 would place the sample at the beginning of the track, and a value of 2.5 would place the sample in the middle of the second measure.
`end_measure`	Decimal value greater than **start_measure** that specifies how long the sample should play for. If the difference between the end and start is greater than the length of the sample, EarSketch will automatically loop the sample for you.

```
fitMedia(RD_UK_HOUSE_MAINBEAT_8, 1, 1, 5)
```

Figure 9.4 Using the fitMedia function in EarSketch.

Here's a quick example of how you would use **fitMedia** to place the sample shown on track 1 in Figure 9.3. You can access the sample from the EarSketch library with a variable named **RD_UK_HOUSE_MAIN-BEAT_8**. We want to place this sample at the beginning of the first track and then extend out for four measures (Figure 9.4).

We'll show more elaborate examples below, but once you have your program entered you can press the **RUN** button to place your code into the DAW for you to play.

9.3.2 The makeBeat function

If you are creating a song based on existing samples, you might want to add your own unique beat. Or maybe you want to add a custom drum fill to transition from one song section to another. The **makeBeat** function creates a drum pattern by specifying the rhythm of a given instrument as a text string. The EarSketch sound library contains a large collection of one-shot percussion instrument samples that you can use with **makeBeat** to create unique drum patterns. The **makeBeat** function has the following arguments:

Parameter	Description
`audio_file_name`	Sample name from the EarSketch library *-or-* string of an audio file name
`track_number`	Integer value greater than or equal to 1 indicating which track to place the beat pattern on
`start_measure`	Decimal value greater than or equal to 1 that specifies the start of the pattern
`rhythm_pattern`	Text string representing the beat pattern. These characters can be used: • "0" plays the sample for a single 16th note • "-" rests for a 16th note • "+" must follow a 0 and extends the sample for another 16th note

The fourth parameter **rhythm_pattern** describes a rhythmic pattern that the audio will follow.

The **rhythm_pattern** string can be any length, but lengths that are multiples of 16 are optimal because they align with the duration of a measure. Here's an example:

```
makeBeat(OS_SNARE03, 2, 1, "0---0+--0+++0000")
```

This takes a sample called **OS_SNARE03** and places a drum pattern on the first measure of the second track. The drum pattern consists of a 16th note on the first beat, an 8th note on the second beat, a quarter note on the 3rd beat, and finally four 16th notes. This is equivalent to the following code in TunePad:

```
playNote(snare, 0.25)
rest(0.75)
playNote(snare, 0.5)
rest(0.5)
playNote(snare, 1)
playNote(snare, 0.25)
playNote(snare, 0.25)
playNote(snare, 0.25)
playNote(snare, 0.25)
```

We can create a basic "four-on-the-floor" drum beat in EarSketch using three lines of code for the kick, snare, and hi-hats.

```
makeBeat(OS_KICK01, 1, 1, "0---0---0---0---")
makeBeat(OS_SNARE01, 2, 1, "----0-------0---")
makeBeat(OS_CLOSEDHAT01, 3, 1, "0+0+0+0+0+0+0+0+")
```

Notice that because the instruments that make up our rhythm overlap with one another, we have to put each one on a separate track.

9.3.3 EarSketch effects

Once you get the hang of adding samples to the DAW, there are many audio effects that you can apply to tracks using a function called **setEffect**. The most basic form of this function applies an effect to the full length of a given track. For example:

```
# Apply a delay effect on track 1
setEffect(1, DELAY, DELAY_TIME, 250)
```

More advanced versions of **setEffect** create dynamic effects that start and end at specific times. For now, here are the basic parameters. You can read the full documentation on the EarSketch site.

Parameter	Description
track_number	Integer value greater than or equal to 1 that specifies which track to apply the effect to. You can also use the **MIX_TRACK** constant to apply the effect to all tracks.
effectType	A constant that specifies the type of the effect. It can be one of the following values: BANDPASS CHORUS COMPRESSOR DELAY DISTORTION EQ3BAND FILTER FLANGER PAN PHASER PITCHSHIFT REVERB RINGMOD TREMOLO VOLUME WAH
effectParameter	Which parameter to set for the given effect. See the online EarSketch documentation for more detail.
effectValue	A decimal (float) number that sets the value for the specified parameter.

9.4 Looping in EarSketch

The fact that we have to specify the start and end measures of a sample in EarSketch creates an interesting problem when we want to repeat a sample multiple times over the duration of a song. For example, suppose you wanted to play a two-beat sample once every measure. How can we do that elegantly with Python code? One solution is to use the loop variable in a Python for loop. The loop variable gets updated every iteration of a loop with the next number in the sequence. For example, this snippet of code would print out the numbers 2, 3, 4, 5:

```
for v in range(2, 6):
    print(v)
```

Notice that our loop variable, **v**, starts at 2 and counts up to (but not in‑cluding) the second parameter of the **range** function. Using this feature of the loop variable, we can get our EarSketch example working with just two lines of code:

```
1  for measure in range(1, 5):
2      fitMedia(SAMPLE, 1, measure, measure + 0.5)
```

In the first iteration through this loop, the value of **measure** is 1, meaning our sample starts on measure 1 and ends at measure 1.5. In the next iteration, the value of **measure** is 2 and our sample begins on measure 2 and ends at measure 2.5. The pattern continues until our loop has completed iterating.

What if we wanted to change our code so that the sample only gets played every other measure? This would mean that we'd somehow work out a way to count by twos instead of ones. The **range** function has an optional third parameter that specifies the amount that our counter in‑creases by each iteration of the loop, which makes this kind of task easy. In the example below, we set up a variable called **skip_count** that we can use to try different numbers (e.g., counting by threes instead of twos). The **skip_count** variable is used as the third parameter in our call to the **range** function on line 2:

```
1  skip_count = 2
2  for measure in range(1, 5, skip_count):
3      fitMedia(SAMPLE, 1, measure, measure + 0.5)
```

9.5 Programming with song structure

Now that we have a rough overview of what we can do in EarSketch, let's think a little bit about how we can create full-length songs with different sections and structures (such as intro-verse-chorus-bridge). In the last few chapters we've talked about how to define our own functions in Python as a way to create reusable segments of code. User-defined functions can also be a great way to think about song structure.

The first step is to pull together material for our song. The duration of our sections will vary, but generally it will be a multiple of four measures. Let's say that we want to create a piece of music that is in ABA form. The

A section starts our piece, followed by a contrasting B section, and ending with a return to the A section. Let's use our "four-on-the-floor" drum pattern along with two vocal loops and a couple of basslines. Here's the full list of samples:

- Kick drum   OS_KICK01
- Hi-Hat   OS_CLOSEDHAT01
- Snare drum   OS_SNARE01
- Vocal lead   ENTREP_VOX_BK_FALSETTO
- Vocal harmony   CIARA_SET_VOX_HOOK_HARMONY
- Bass line 1   ELECTRO_ANALOGUE_BASS_001
- Bass line 2   ELECTRO_ANALOGUE_BASS_009

Now that we have the materials for our song, we can assemble them into the ABA pattern. Here's the full EarSketch script.

```
from earsketch import *

init()
setTempo(90)

### section A -4 bars (1–4)
for bar in range(1, 5):
    makeBeat(OS_KICK01, 1, bar, "0---0---0---0---")
    makeBeat(OS_SNARE01, 2, bar, '----0-------0---')
    makeBeat(OS_CLOSEDHAT01, 3, bar, '0+0+0+0+0+0+0+0+')
    fitMedia(CIARA_SET_VOX_HOOK_HARMONY, 4, bar, bar+1) # vocal loop

for bar in range(1, 5, 2):
    fitMedia(ELECTRO_ANALOGUE_BASS_001, 5, bar, bar+2) # bass


### section B -4 bars (5–8)
for bar in range(5, 9):
    makeBeat(OS_KICK01, 1, bar, "0---0---0---0---")
    makeBeat(OS_SNARE01, 2, bar, '----0-------0---')
    makeBeat(OS_CLOSEDHAT01, 3, bar, '0+0+0+0+0+0+0+00')
    fitMedia(ENTREP_VOX_BK_FALSETTO, 4, bar, bar+1) # vocal loop

for bar in range(5, 9, 2):
    fitMedia(ELECTRO_ANALOGUE_BASS_009, 5, bar, bar+2) # bass


### section A -4 bars (9–13)
for bar in range(9, 13):
    makeBeat(OS_KICK01, 1, bar, "0---0---0---0---")
    makeBeat(OS_SNARE01, 2, bar, '----0-------0---')
    makeBeat(OS_CLOSEDHAT01, 3, bar, '0+0+0+0+0+0+0+0+')
    fitMedia(CIARA_SET_VOX_HOOK_HARMONY, 4, bar, bar+1) # vocal loop
```

```
for bar in range(9, 13, 2):
    fitMedia(ELECTRO_ANALOGUE_BASS_001, 5, bar, bar+2) # bass

finish()
```

Notice that, when the A section reappears, the code is almost exactly the same—the only difference is the start and end measures of our sections. The rest of the code remains unchanged. Whenever you see large amounts of repeated code, you should be thinking about creating either a function or a loop. Remember your goal as a programmer is to create code that is as simple and elegant as possible. Let's try writing functions that specify the start and duration values as **parameters**. The loop inside of the functions iterates from the **start** value to **start + duration**, for a total of **duration** iterations. Here's a revised version of the code using functions:

```
from earsketch import *

init()
setTempo(90)

def section_a(start, duration):
    for bar in range(start, start+duration):
        makeBeat(OS_KICK01, 1, bar, "0---0---0---0---")
        makeBeat(OS_SNARE01, 2, bar, '----0-------0---')
        makeBeat(OS_CLOSEDHAT01, 3, bar, '0+0+0+0+0+0+0+0+')
        fitMedia(CIARA_SET_VOX_HOOK_HARMONY, 4, bar, bar+1)

    for bar in range(start, start+duration, 2):
        fitMedia(ELECTRO_ANALOGUE_BASS_001, 5, bar, bar+2)

def section_b(start, duration):
    for bar in range(start, start+duration):
        makeBeat(OS_KICK01, 1, bar, "0---0---0---0---")
        makeBeat(OS_SNARE01, 2, bar, '----0-------0---')
        makeBeat(OS_CLOSEDHAT01, 3, bar, '0+0+0+0+0+0+0+00')
        fitMedia(ENTREP_VOX_BK_FALSETTO, 4, bar, bar+1)

    for bar in range(start, start+duration, 2):
        fitMedia(ELECTRO_ANALOGUE_BASS_009, 5, bar, bar+2)
section_a(1, 4)
section_b(5, 4)
section_a(9, 4)

finish()
```

Why is this method of using functions better? It has the same end result as the first version we tried. The difference is that if we want to change a note or rhythm, we only need to change it once. If we want to reorder our section, we just have to change a single function call. If we're trying to debug an error, we can easily isolate the bug in one function. The code with functions is also much more readable, especially at the end where we're ordering our song sections.

9.6 Next steps with EarSketch

EarSketch has a vibrant online community along with extensive documentation and video tutorials. We encourage you to spend some time exploring the platform, trying out some sample projects, and making your own music. This is a good way to apply your Python coding skills in another context or even to try out another language like JavaScript. Have fun and see what you can create!

Interlude 9

HOW TO MAKE A SNARE DRUM RISER

A snare drum riser is an accelerating drum roll that marks a transition point in a song. It's a common technique for adding energy and excitement before the beat drops. In this tutorial you'll code a basic riser pattern in Python and then add in a few effects like randomized volume, pitch shifting, and panning. You can follow along online in TunePad at https://tunepad.com/interlude/snare-riser.

STEP 1: The basic pattern

One simple version of a snare riser plays 4 sets of 8 notes with each new round of notes played twice as fast as the previous round. This creates an accelerating pattern that looks something like this (Figures 9.5).

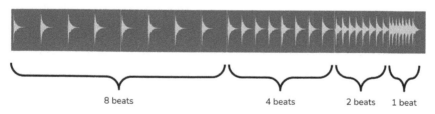

| 8 beats | 4 beats | 2 beats | 1 beat |

Figure 9.5 Snare drum riser pattern.

DOI: 10.4324/9781003033240-18

There are a few different ways we could create this pattern in code. Before you look at what we came up with, spend a few minutes trying to come up with your own solution.

```
1  duration = 1.0              # duration of each note
2  for i in range(4):          # 4 sets of notes
3      for j in range(8):      # each set has 8 notes each
4          playNote(2, beats = duration)
5      duration = duration / 2  # cut the note duration in half
```

Our version uses two for loops, one nested inside of the other. The outer loop repeats four times (line 2), and the inner loop repeats eight times (line 3). This means that the **playNote** on line 4 gets called 8 x 4 = 32 times. The length of each note is determined by the duration variable that we declare on line 1. This variable starts out with a value of 1.0, but in each iteration of the outer loop, we divide the value in half.

Alternative pattern

Here's an alternative snare riser pattern where each note is played slightly faster than the previous note (Figures 9.6).

This pattern only uses one loop, and we reduce the duration variable each go around by multiplying it times 0.94 (making it 94% of its current value). This kind of tapering off is called an exponential decay.

```
1  duration = 1.0
2  for i in range(40):
3      playNote(2, beats = duration)
4      duration = duration * 0.94
```

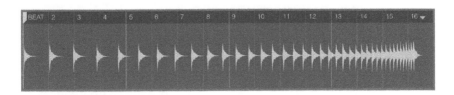

Figure 9.6 Alternative snare drum riser pattern.

STEP 2: Randomized volume

Now that we have our basic patterns, we can start to add some different effects. First, let's create a more human-like feel by adding variation in the volume of each snare drum hit. We can do this with Python's built-in **randint** function. Calling **randint** generates a random integer number between two values. In the example below, we generate a random value between 85 and 100 that we pass into the **velocity** parameter of **play-Note**. The effect is a subtle but noticeable variation in the texture of the sound.

```
1   from random import randint
2
3   duration = 1.0
4   for i in range(4):                 # four sets of eight beats
5       for j in range(8):
6           volume = randint(85, 100)  # generate a random number
7           playNote(note = 2,         # play the snare drum sound
8                    beats = duration, # duration of each note
9                    velocity = volume) # randomized volume
10          duration = duration / 2    # cut the note duration in half
```

STEP 3: Rising pitch

To build a sense of rising excitement, it helps to increase the pitch of the snare drums across the run. We can use the **with bend** instruction in TunePad to achieve this effect. We provide the effect with a number measured in **cents** that says how much to change the pitch. For example, a value of 100 cents increases the pitch by one semitone (the distance between a C and a C#). And, a value of 200 cents increases the pitch by two semitones (the distance between a C and a D). We're going to shift the pitch by 1,000 cents over a period of 16 beats to increase 10 semitones.

```
1   from random import randint
2
3   with bend(cents = 1000, beats = 16):
4       for i in range(4):
5           for j in range(8):
6               playNote(note = 2,
7                        beats = pow(2, -i),
8                        velocity = randint(85, 100))
```

In this version of the code, we've simplified things a little by getting rid of the **duration** and **volume** variables. Instead, we perform the same calculations directly in-line when we call **playNote**.

STEP 4: Putting it all together

To finish up, here's a ready-to-use **snare_riser** function that wraps the code inside a single function that you can use in your next project. The **def** keyword on line 1 is used to *declare* a function that we've named **snare_riser**. After you declare the function, you can use it by calling it by name: **snare_riser().**

```
1   from random import randint
2
3   def snare_riser():
4       for i in range(4):
5           for j in range(8):
6               playNote(note = 2,
7                   beats = pow(2, -i),
8                   velocity = randint(85, 100))
```

Optional: Pan widening

Here's one final effect that uses two different snare drum sounds played simultaneously. The first is panned hard left and the second hard right to widen out the sound. That means we hear one sound entirely out of the left speaker and the other entirely out of the right speaker. Here's an updated version of our **snare_riser** function that includes this effect. The secret ingredients are the **with pan** effect that takes a number between -1 and 1 that specifies where in the stereo spectrum the sound should come from (-1.0 means all the way left, 1.0 means all the way right, and 0.0 means centered). We also use the **rewind** function in TunePad to play both the left and right snare sounds simultaneously. If we didn't use rewind, the sounds would play one after another.

```
1    from random import randint
2
3    def snare_riser():
4        with bend(cents = 1000, beats = 16):
5            for i in range(4):
6                for j in range(8):
7                    with pan(-1):
8                        playNote(note = 2,
9                            beats = pow(2, -i),
10                           velocity = randint(85, 100))
11                   # rewind so that both notes play at the same time
12                   rewind(pow(2, -i))
13                   with pan(1):
14                       playNote(note = 3,
15                           beats = pow(2, -i),
16                           velocity = randint(85, 100))
```

Going even farther

There are many other effects we could add to this example. Some common effects for a riser:

• Lowpass filter sweep that opens up the frequency spectrum as the riser accelerates
• Other effects like flangers or phasers
• Adding swing or other subtle variations in timing
• Using tonal instruments or other percussion sounds instead of a snare drum
• Variations in the basic patterns that we've introduce here

As you experiment with different variations, create new functions so that you can easily drop them into your next project.

10 Modular synthesis

Guest chapter by Dillon Hall

Modular synthesis is a set of tools and techniques for electronically or digitally "synthesizing" musical sounds. *Modular* means "made up of smaller pieces that can be taken apart and rearranged". *Synthesis* means "creating something new from existing parts or ideas". Modular synthesizers were first introduced in the 1950s and 1960s with devices like the Moog System 55 (Figure 10.1), which were sometimes the size of entire rooms. These synthesizers were *analog*, meaning they generated sounds using electrical circuits instead of starting with some kind of physical vibration (like a guitar string). By chaining together electronically generated waveforms with filters, effects, and envelopes, they could not only approximate traditional acoustic instruments—like bass, piano, brass, woodwinds, and drums—but also create entirely new sounds altogether. Today, we can make these same sounds on a computer using *digital signal processing* techniques. While the process has changed, the basic ideas are the same: take several different **modules** and combine them together into a **patch**. A module is a single device with a single purpose such as creating, transforming, or controlling sound. A patch is formed by chaining many modules together to combine sounds and layer effects. This chapter provides a high-level overview of modular synthesis concepts using an interface built into TunePad.

DOI: 10.4324/9781003033240-19

Figure 10.1 Moog System 55 modular synthesizer.

10.1 Signals

In modular synthesis, a **signal** is a collection of values that varies over time and can carry information. In the physical world, when sound waves collide with the membrane of a microphone, they cause fluctuations in voltage levels of an electric circuit. The microphone has transformed physical sound waves into an electric signal that consists of fluctuating voltage levels. These recorded voltage levels are just one example of a signal (Figure 10.2).

For analog synthesizers, this process works in reverse. Audio signals are generated synthetically using simple electronic components—such as resistors, capacitors, and transistors—rather than a microphone. By varying the voltage input levels, we can change the frequency at which the circuit oscillates or "vibrates". In the digital world, instead of using oscillator circuits, computers generate streams of numbers that simulate the voltage output of the original electronic components.

For modular synthesis, there are two basic kinds of signals: *audio signals* and *control signals*. Audio signals oscillate in the range of human hearing, meaning that they can create musical notes. The faster they move, the higher their *frequency* and the higher the pitch we hear. Control signals, on the other hand, tend to vibrate more slowly, below the range of human hearing. Instead of sending them directly to our speakers, we use them to change parameters of audio signals. A good example of this is vibrato. A violinist draws a bow across a string to generate a high-frequency sound that falls in the range of human hearing. As the bow is drawn, the violinist

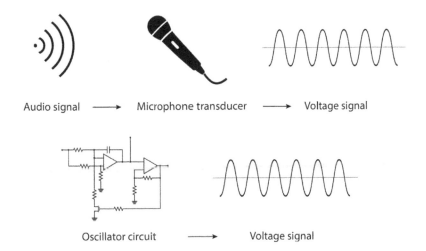

Audio signal ⟶ Microphone transducer ⟶ Voltage signal

Oscillator circuit ⟶ Voltage signal

Figure 10.2 Audio signal from a microphone (top) and an audio signal from an electric circuit.

The oscillator circuit in Figure 10.2 is modified from a drawing by Krishnavedala (commons.wikimedia.org/wiki/User:Krishnavedala). Creative Commons License creativecommons.org/licenses/by-sa/4.0/.

quickly rocks a finger on the fretboard to modulate the pitch. The violinist's finger operates like a control signal that modifies the audio signal of the violin string.

10.2 Modules

One way to think about modules is as physical pieces of equipment with one or more sockets for inputs and an output signal. Inputs may be audio signals or control signals. Outputs may feed into other modules or be sent to our speakers. In the simple TunePad patch below, the sine wave module generates an audio signal that gets plugged into the Output module and sent to the speakers (Figure 10.3).

sine 1   output

Figure 10.3 A simple Modular Synthesis patch created in TunePad.

10.2.1 Source modules

A source module is a type of module that generates a signal. Sources can also have inputs to control parameters like pitch and amplitude.

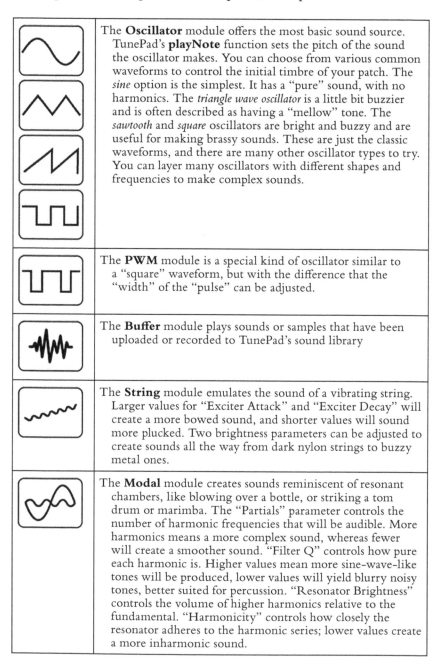

	The **Oscillator** module offers the most basic sound source. TunePad's **playNote** function sets the pitch of the sound the oscillator makes. You can choose from various common waveforms to control the initial timbre of your patch. The *sine* option is the simplest. It has a "pure" sound, with no harmonics. The *triangle wave oscillator* is a little bit buzzier and is often described as having a "mellow" tone. The *sawtooth* and *square* oscillators are bright and buzzy and are useful for making brassy sounds. These are just the classic waveforms, and there are many other oscillator types to try. You can layer many oscillators with different shapes and frequencies to make complex sounds.
	The **PWM** module is a special kind of oscillator similar to a "square" waveform, but with the difference that the "width" of the "pulse" can be adjusted.
	The **Buffer** module plays sounds or samples that have been uploaded or recorded to TunePad's sound library
	The **String** module emulates the sound of a vibrating string. Larger values for "Exciter Attack" and "Exciter Decay" will create a more bowed sound, and shorter values will sound more plucked. Two brightness parameters can be adjusted to create sounds all the way from dark nylon strings to buzzy metal ones.
	The **Modal** module creates sounds reminiscent of resonant chambers, like blowing over a bottle, or striking a tom drum or marimba. The "Partials" parameter controls the number of harmonic frequencies that will be audible. More harmonics means a more complex sound, whereas fewer will create a smoother sound. "Filter Q" controls how pure each harmonic is. Higher values mean more sine-wave-like tones will be produced, lower values will yield blurry noisy tones, better suited for percussion. "Resonator Brightness" controls the volume of higher harmonics relative to the fundamental. "Harmonicity" controls how closely the resonator adheres to the harmonic series; lower values create a more inharmonic sound.

10.2.2 Control modules

Control modules generate signals that aren't typically audible to humans. Control modules can feed into the input sockets of other modules.

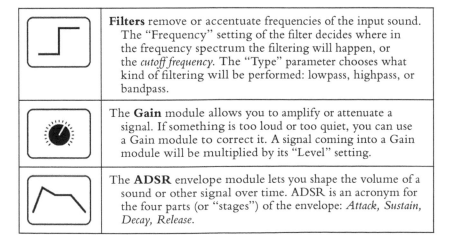	The **Constant Signal** module is the simplest control signal. It outputs an unchanging signal with an adjustable value.
LFO	The Low Frequency Oscillator or **LFO** module works the same as the Oscillator module used to generate signals, but it moves much more slowly—hence the "low frequency". LFOs are primarily used to change the parameters of other modules. For instance, you could create a vibrato effect, which is the slight fluctuation of pitch, by using an LFO to **detune** an oscillator. The LFO has a "waveform" parameter similar to the regular oscillator. "Sine" and "Triangle" will move parameters up and down in a smooth and natural way. "Square" will jump between high and low. "Sawtooth" will ramp up slowly and then suddenly drop.
	The **Random Signal** module outputs a stream of random numbers within a specified range. This module can be used to add uncertainty to a synth patch, or to provide the kind of small variation between notes that is present when playing an acoustic instrument. The random signal changes every time a new note is played.

10.2.3 Processor modules

Processor modules alter input signals. Many of these are the same effects we might use for mixing: filters, gain, reverb, echo, compression, and ADSR envelope. These can be used to give your patch more spice and to sculpt your desired sound.

	Filters remove or accentuate frequencies of the input sound. The "Frequency" setting of the filter decides where in the frequency spectrum the filtering will happen, or the *cutoff frequency*. The "Type" parameter chooses what kind of filtering will be performed: lowpass, highpass, or bandpass.
	The **Gain** module allows you to amplify or attenuate a signal. If something is too loud or too quiet, you can use a Gain module to correct it. A signal coming into a Gain module will be multiplied by its "Level" setting.
	The **ADSR** envelope module lets you shape the volume of a sound or other signal over time. ADSR is an acronym for the four parts (or "stages") of the envelope: *Attack, Sustain, Decay, Release.*

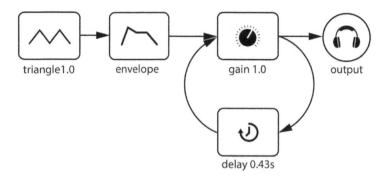

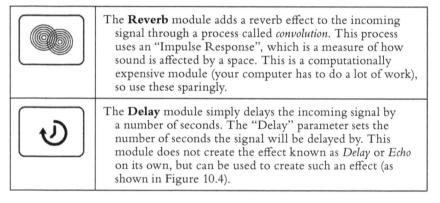

Figure 10.4 A patch with a delay effect.

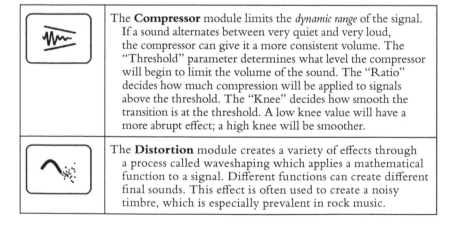

(icon)	The **Reverb** module adds a reverb effect to the incoming signal through a process called *convolution*. This process uses an "Impulse Response", which is a measure of how sound is affected by a space. This is a computationally expensive module (your computer has to do a lot of work), so use these sparingly.
(icon)	The **Delay** module simply delays the incoming signal by a number of seconds. The "Delay" parameter sets the number of seconds the signal will be delayed by. This module does not create the effect known as *Delay* or *Echo* on its own, but can be used to create such an effect (as shown in Figure 10.4).

In the patch shown in Figure 10.4, the Gain module allows the sound to pass through unchanged, while also routing the signal to the Delay module. The Delay module has "Delay" set to 0.43 seconds, and has its "Level" parameter set *less than 1*. This is important to prevent the signal from getting louder and louder with every echo.

(icon)	The **Compressor** module limits the *dynamic range* of the signal. If a sound alternates between very quiet and very loud, the compressor can give it a more consistent volume. The "Threshold" parameter determines what level the compressor will begin to limit the volume of the sound. The "Ratio" decides how much compression will be applied to signals above the threshold. The "Knee" decides how smooth the transition is at the threshold. A low knee value will have a more abrupt effect; a high knee will be smoother.
(icon)	The **Distortion** module creates a variety of effects through a process called waveshaping which applies a mathematical function to a signal. Different functions can create different final sounds. This effect is often used to create a noisy timbre, which is especially prevalent in rock music.

The **Bitcrusher** module emulates the sound of older digital music production tools. This effect essentially creates distortion by discarding information from an input signal. Lower values of sample rate and bit depth will create more pronounced effects. Early sampled music, like 90s hip-hop and dance music, had a distinct low-sample-rate sound that can be emulated by using the Bitcrusher with a low-sample-rate value. Lower values of "bit depth" will create sounds similar to those created on video game consoles and can be used to make chip-tune sounds.

The **Stereo Merger** node allows you to decide exactly what the left and right audio channels will sound like. Anything connected to the "left" will go into the left audio channel, and vice versa for the "right". The output of the Stereo Merger is a single signal that otherwise works the same as mono signals within the patch designer.

The **Mixer** module blends two sounds together. You can use the "Mix" parameter to choose how much of each signal to allow trough. Leaving "Mix" at 0 and connecting an LFO will smoothly move back and forth between both sounds.

10.3 Creating patches

Now that we have seen several different modules, let's build a first patch. Go to the TunePad sound designer https://tunepad.com/sound-design and create a new patch. You should see a blank patch that consists of only an Output module. You can move that module around by clicking and dragging it. There will also be a menu of modules to choose from. Let's start by adding a sine wave oscillator. Now click and drag starting with the small arrow on the right side of the Oscillator module. A line will appear that you can drag to connect to the left side of the output module (see Figure 10.3).

Congrats! You've made your first patch with the TunePad instrument designer. Try playing the piano keys to hear how it sounds. In order to improve this patch you can first edit the parameters of each module and listen to the effects. The next section covers strategies for synthesizing a variety of different sounds.

Many of the modules have a "Level" control. Your first step in creating a patch should always be setting the "Level" controls of your modules. Start by lowering the Output's level to 0.5 or so and playing some notes. Designing musical instruments is fun, but it's important to protect your hearing. *Start with your volume low and adjust it as needed!*

10.4 Synthesis techniques and algorithms

Now that we have a grasp on signals and have seen a variety of common modules, let's take a look at common strategies for combining these concepts to make our own digital instruments.

10.4.1 Additive

Additive Synthesis is a technique that *adds* multiple audio signals together to make more complex sounds. This technique is inspired by the process of *spectral analysis* in which a complex signal is broken down into the individual frequencies that make it up. Since sine waves contain only one frequency, combining them is essentially spectral analysis in reverse. A simple Additive Synthesis patch might replicate the sound of an organ by combining several sine waves at integer multiples of a fundamental frequency. More oscillators create a more complex timbre.

Here four different sine waves are added together (Figure 10.5, left). The top sine wave has a "Frequency Multiple" of 1, meaning it is the *fundamental* frequency. The other oscillators have "Frequency Multiples" of 2, 3, and 4, and function to add a more complex timbre. Make sure to turn down the "Level" of your oscillators so they are not too loud! An easy variation on the last patch is to add different envelopes to different oscillators (Figure 10.5, right). In this patch the higher oscillators have a longer "Attack Time" and will thus take longer to fade in. Different acoustic instruments have different changes in harmonics over time that can be recreated with Additive Synthesis.

10.4.2 Subtractive

The most popular and widely used synthesis technique is ***Subtractive Synthesis***. Whereas Additive Synthesis layers harmonics onto a sound through the use of multiple oscillators, Subtractive Synthesis starts with these harmonics and strips them away. This method employs buzzy oscillators, filters, and envelopes to create sounds ranging from organic to other-worldly (Figure 10.6).

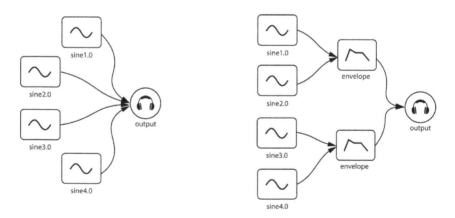

Figure 10.5 Two variations of an Additive Synthesis patch.

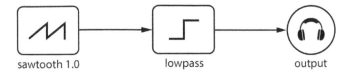

sawtooth 1.0   lowpass   output

Figure 10.6 A simple Subtractive Synthesis patch.

A simple Subtractive Synthesis patch might feature a sawtooth oscillator attached to a lowpass filter. The sawtooth oscillator waveform is very
bright and has lots of harmonics for the filter to interact with.

10.4.3 FM Synthesis

The "FM" in **FM Synthesis** stands for "Frequency Modulation". The idea
is to modulate the frequency of an oscillator over time. When done slowly,
this sounds like a vibrato effect, but when done quickly it can create very
interesting changes to the timbre of a sound (Figure 10.7).

The simplest FM patch is pictured above. Notice that the left-most
oscillator is not connected to the audio input, but rather the "frequency"
input of the second oscillator. We will not hear the output of the first
oscillator, and are instead using it to change the sound of the second
oscillator. We can refer to oscillators like the first one as *modulators*.
Modulators exist to change how other oscillators sound, but we don't
end up hearing them. The second oscillator is being modulated by the
first, and sends its output as audio to the Output module. An oscillator
that receives modulation like this is called a *carrier*. The degree to which
the FM effect takes place is up to the "Frequency Mod" parameter on
the carrier oscillator.

As with other synthesis techniques, we can set up a patch that changes
the timbre of the sound over time with various methods of modulation.
The following patch demonstrates how to do this with an ADSR or an
LFO. The chain of modules on top are affected by the LFO on the left. This

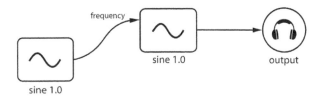

frequency

sine 1.0   output

sine 1.0

Figure 10.7 An FM Synthesis patch where one sine wave is fed into the frequency
input socket of another sine wave.

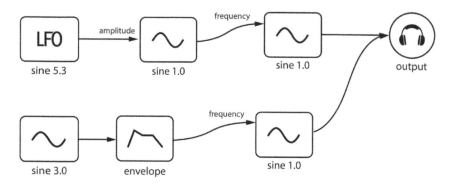

Figure 10.8 A more complex FM Synthesis patch.

LFO is connected to the "amplitude" input on the modulator oscillator to its right. Thus, the carrier oscillator is moving slowly between a bright complex tone and a simple sine wave according to the LFO (Figure 10.8).

The lower chain of modules is affected by an ADSR. The modulator oscillator to the left will shape the sound of the modulator, but only as the ADSR permits. This means that the sound will be brightest when the ADSR allows the modulator to have the highest amplitude. The carrier oscillator will move from a pure tone into a brighter one as the Attack ramps up to maximum amplitude, and will have a simpler, darker tone as the ADSR brings the modulator down to the sustain amplitude and eventually to silence in the release stage.

10.4.4 Physical modeling

Physical Modeling Synthesis refers to many techniques that attempt to mimic the physical processes by which sounds are produced acoustically. The String and Modal modules perform physical modeling algorithms on their own, but the Delay, Filter, and Reverb modules can be used to mimic other acoustic phenomena. The two most common elements to physical modeling synthesis are *excitation* and *resonance*. The Modal module, for example, has an *exciter*, which is a burst of white noise, and a *resonator* which is a collection of bandpass filters. The exciter can be thought of as a mallet striking a surface, or someone blowing over a bottle. This sound is processed by the resonator, whose bandpass filters will isolate certain frequencies to give a variety of sonic possibilities.

The patch pictured below is an example of how different physical modeling algorithms can be combined to make interesting instruments. Here the output of the String module is sent to a Modal module, which has its exciter turned off. The Modal module is used as an additional resonator, on top of the Karplus–Strong resonator inside the String module. The Modal

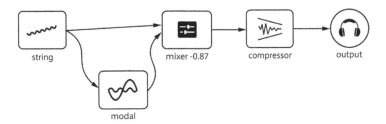

Figure 10.9 An example of physical modeling to create complex string-like sounds.

resonator tends to sound like a hollow box, so you could imagine this as an instrument with strings vibrating over a wooden chamber, like an acoustic guitar, koto, or cello (Figure 10.9).

10.5 Examples

Here are a few more examples to get you started designing your own patches.

10.5.1 Example 1: Basic poly/lead

This patch is a popular and flexible sound that uses *Subtractive Synthesis*. This type of patch is good as a lead for playing melodies, and also as a **polyphonic** or **"poly" synth**, putting a chord progression at the front of a mix without being too muddy. A polyphonic synth is a synth capable of playing more than one note at a time and is well suited to carry the harmony. First let's set the envelope of our Output node to be something quick and responsive. The left column of the table below shows the values:

Output ADSR		Sawtooth detune		Filter ADSR		Filter parameters	
Attack	0.1	Detune 1	0	Attack	0.12	Type	lowpass
Decay	0.0	Detune 2	−3	Decay	0.1	Frequency	350 Hz
Sustain	1.0	Detune 3	5	Sustain	0.13	Frequency Mod	75%
Release	0.2			Release	0.2		

Now for the oscillators, let's try stacking several sawtooth oscillators that are slightly out of tune with one another to create a "super saw" (far left side of Figure 10.10). A variation on this patch could use square or triangle oscillators.

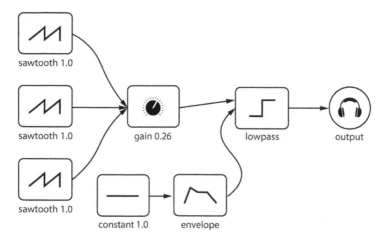

Figure 10.10 Complete Basic Poly/Lead patch in TunePad.

To achieve the "super saw" effect the detune parameter is set to 0, −3, and 5, respectively, on each oscillator. Notice that the oscillators are connected to a Gain node to create a hub for our oscillator section, and attenuate the volume of all our oscillators at a single point. The Level parameter is set to -15 dB.

If you play a few notes you should hear a familiar buzzy sawtooth sound. You could stop here and you'd have a capable lead patch, but let's keep going and add some dynamics to this sound. To do this we'll add a filter, an ADSR, and a Constant Signal to power the ADSR (bottom middle of Figure 10.10). The envelope will allow us to open and close the filter as notes are played, which will give our patch depth and realism. Connect the envelope to the frequency input of the filter. The ADSR connected to the Filter (the "Filter Envelope") has settings in the table above.

The patch should have a punchy bright attack that cools off into a mellow sustain, because the Filter Envelope quickly opens and closes the filter on every key press. This is perfect for a balanced mix because the attack catches your attention and lets you know the note is there, but the sustain loses the higher frequencies and makes room for the other instruments in your track. This is why this patch is so flexible and useful as both a lead and a poly.

10.5.2 Example 2: FM Arp Bell

This is a plucky patch, perfect for chord arpeggios with a distinct bell-like sound reflective of its use of *FM Synthesis*. To start we need oscillators; for FM it's common to start with sine oscillators. Wire them up as shown in Figure 10.11. The oscillator on the left is our *modulator* and the one in the

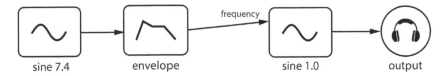

Figure 10.11 Complete FM Synthesis example.

middle is the *carrier*. Set the "Frequency Mod" of the carrier oscillator to somewhere around 80%, for a well-pronounced effect. Since FM sounds doen't always sound the same across the keyboard, use a higher octave to test it out.

To make a bell sound we want to find a good ratio of frequencies that sound *inharmonic* without sounding ugly. The closer the ratio is to a whole number the more the harmonic. A modulator at 2 and carrier at 1 is *much more* harmonic than 2.1 to 1. A higher ratio of frequencies results in higher harmonics in the resulting sound. If this number is too high, the frequencies will begin to fall outside the range of human hearing. A ratio of 7.4 to 1 sounds sufficiently metallic, without sounding out-of-this-world. Now that we have our oscillators set up it's time to put some envelopes in place. To get a good bell-like envelope, the Output envelope should be set as shown in the table:

Output ADSR		Carrier ADSR		Sine frequencies	
Attack Time	0.01	Attack Time	0.01	Modulator	7.4
Decay Time	0.65	Decay Time	0.4	Carrier Frequency Mod	80%
Sustain	0.0	Sustain	0.0		
Release Time	0.25	Release Time	0.15		

At this point our bell sound is sounding pretty good, but there is no timbral change over time, only a change in volume. We can add another envelope between the modulator and the carrier to create a sound that starts off bright and inharmonic, then decays into a sine wave. This is very similar to how the harmonics of real bells change over time. Set the carrier envelope ADSR parameters as shown in the table above.

10.6 Conclusion

The world of modular synthesis is populated with enthusiastic hobbyists and professionals who work in both hardware and software, and there are many online resources to learn more. This chapter has reviewed the basics of signals and tools used to process signals to generate sounds. The next chapter reviews some of the history of synthesizers and how they relate to computer-generated music.

Interlude 10

DRUM MACHINE FUNCTION

Overview

A common way to create drum patterns and rhythms is to use a step sequencer. A step sequencer breaks up time into even slices or steps (like 16th notes or 32nd notes). This Roland 808 drum machine from the 1980s let musicians program rhythms with 16 switches representing times when various drum sounds would be played. Modern step sequencers use grids where each row is a different drum sound and each column is a consecutive time slice. In this interlude we're going to make a custom drum machine in TunePad with Python code. You can follow along online with this TunePad project

  https://tunepad.com/interlude/drum-machine.

STEP 1: Machine function

We're going to make a drum machine in Python using our own function called **machine**. Our function will use **strings** to make sound patterns for each drum. We'll also need a TunePad function called **moveTo** that lets us fast forward or rewind time by moving the beat counter to any position we want. Start by creating a new Drum instrument in a project and enter this code:

DOI: 10.4324/9781003033240-20

```
1  def machine(note, pattern):
2      moveTo(0)
3      for s in pattern:
4          if s == '!':
5              playNote(note, beats=0.25, velocity=100)
6          else:
7              rest(0.25)
```

This function is only seven lines long, but there's a lot going on. Let's break it into easier pieces.

- First, the function takes two parameters, *note* and *pattern*. The **note** parameter is just a number representing the drum sound to play (like 0 for kick, 2 for snare, or 4 for hat). The **pattern** parameter is something called a **string** in Python. We can use strings to contain combinations of characters. For example, in this string, **"Hello, I'm a string"**, there are 18 characters starting with **"H"** and ending with **"g"** and including the spaces. Python uses either double quotes (") or single quotes (') to define a string.
- The first line of the function uses **moveTo** to move back to the beginning of the measure. We'll see why this is important when we make multiple calls to **machine** (below).
- Next, the machine function uses a for loop to iterate over each character in the string, one at a time. This is slightly different from the counting loop that we used for hi-hats. This loop **iterates** over a list of *characters* instead of a list of numbers.
- Line 4 has a statement called a **conditional**, or an **if-then-else statement**. What we're trying to do is say *if* the character is a "!", play a sound really loud (velocity = 100). Otherwise, rest. In a minute, we'll revisit this to also let us play quiet beats. Notice again that we use the colon character and indentation to set up blocks of code.

STEP 2: Using the machine function

OK, but how do we use this function? Here's a simple example beat with five sounds. Each call to **machine** is like a row in the drum machine. We use the "-" character to mean no sound, but really this can be any character except at "!".

```
machine(8,  "------!!---------")   # tom 1
machine(6,  "-------------!-!-")   # tom 2
machine(10, "-------!!!------")    # clap
machine(4,  "-!!!-!!!-!!!-!!!")    # hat
machine(2,  "----!----!!-!---")    # snare
machine(0,  "!!------!!!!----")    # kick
```

STEP 3: Loud and soft drum sounds

So, how about both loud and soft sounds to add a little texture to our beat. For that, we add a new statement called **elif** that stands for ("else if") or ("otherwise if"). This is the same function as before, but we're redefining it to recognize characters as different volume indicators. Try replacing your original **machine** function with this code.

```
1  def machine(note, pattern):
2      moveTo(0)
3      for s in pattern:
4          if s == '!':
5              playNote(note, beats=0.25, velocity=100)
6          elif s == '*':
7              playNote(note, beats=0.25, velocity=80)
8          else:
9              rest(0.25)
```

Here's one last version of our drum beat with both loud and soft notes. Press the play button to hear how it sounds. Try changing the Python strings to create different beat patterns.

```
machine(8,  "------!*--------")        # tom 1
machine(6,  "-------------!-")         # tom 2
machine(10, "-------!*!*-----")        # clap
machine(4,  "-*!*-*!*-*!*-*!*")        # hat
machine(2,  "----!----**-!---")        # snare
machine(0,  "!*------!*!*----")        # kick
```

11 History of music and computing

Guest chapter by Jason Freeman

Throughout this book, you've seen that music and computing are connected. Musicians think about things like notes, meter, and phrases just
like programmers think about things like variables, functions, and loops.
Code can help us understand the inner logic of how music is created.
Writing music with code can help us create interesting music and express
ourselves in new ways.

These connections between music and computing are no accident.
Since the earliest computers, musicians have been inventing ways to use
computers in music. Today, when you listen to your favorite song, computers have likely been involved in many different ways. For example:

- The songwriter may have used music notation or audio recording
  software to capture their initial inspiration for the song on a laptop,
  tablet, or mobile phone.
- Musicians in the band may have performed on digital keyboards,
  drum machines, or control surfaces. Guitarists and bass players may
  have routed their audio through digital effects pedals. Each of these
  devices makes or transforms sound through an embedded computer.
- In the recording studio, a producer and audio engineer probably recorded, edited, and mixed the song using digital audio workstation
  software on a computer. They may have even used additional software
  plugins for special tasks like eliminating background noise or fixing
  notes that were out of tune.

DOI: 10.4324/9781003033240-21

- A music streaming service had to store the music track and information about it in a cloud-based database. The service also developed an app to stream the track to your device. In addition, the service probably developed machine-learning algorithms to generate personalized playlists or radio stations for you.

In this chapter, we will explore some of the key moments in the history of computer music. This history will help us understand how computing is continuing to revolutionize the creation, performance, and distribution of music today.

11.1 What is computer music?

Computer music refers to any process of creating, performing, analyzing, or distributing music that involves a computer. This includes a wide range of activities, from writing code to creating music (like in EarSketch and TunePad) to producing new music with an app to watching your favorite music videos online.

Because computer music is such a broad field, it helps to ask a few basic questions as we look at developments in computer music history or think about the things we want to do with computers ourselves:

What is the role of the computer? We often think of computers as tools that help us get a job done, like writing a paper or reading a message. Computers can be musical tools too. They can be musical instruments that create sound in response to our actions, like a digital keyboard that plays a note every time we press a key. Most of the instruments in TunePad, for example, work in this way. We press keys on the computer keyboard and immediately hear the corresponding instrument sound from TunePad.

Computers can also act more like a musician who performs a piece, taking each note in a score and turning it into sound. Much of the code we write in TunePad works this way. We specify a series of notes to be played. Then TunePad displays the notes in a grid and plays them back for us. Each instrument in a TunePad dropbook acts like a single musician. When we play them all back together, the computer becomes like a band of musicians.

Many computer musicians also explore how computers can act like intelligent musicians. These intelligent musicians don't simply follow the instructions in a score. An artificially intelligent computer composer may create its own musical scores that mimic the style of a human musician. An artificially intelligent computer performer may listen to human musicians and improvise along with them.

How does the computer represent music? Human musicians have many different ways of representing music: we create notated scores, we create shorthand lead sheets, we talk about it in words, and we teach it by oral tradition. Computer musicians also have many different ways of representing music. Two approaches are most common. Symbolic representations describe music as a series of musical notes, each with a specific pitch, loudness, start time, and duration. Audio representations describe music in terms of the actual sound that we hear.

You have already seen these two approaches with TunePad and EarSketch. TunePad's **playNote()** function, for example, is a symbolic approach that defines the properties of a single musical note. Music in TunePad consists of a bunch of these notes defined by playNote() and other TunePad functions.

In contrast, EarSketch's **fitMedia()** function adds an audio file into your song. That audio file can be anything: it can be a single musical note, it can be an entire musical phrase, or it can be a sound effect that might not even seem like music. A song in EarSketch is made up of a bunch of these audio files placed on a multi-track timeline.

There are advantages and disadvantages to each approach. With a symbolic representation, the computer can easily represent and manipulate each individual note. TunePad can easily draw the music on a grid that shows each note, and if you want to change a single note in your song, it is easy to find the corresponding line of code and update it. With an audio representation, those kinds of visualizations and edits are often not possible. But there is more flexibility to capture, generate, and edit any kind of sound.

Does the computer create music in real time? Early computers often took hours to generate a few seconds of sound, and even today, computer music applications that use machine learning and artificial intelligence (AI) can take days to analyze large datasets. Many computer music applications, like musical instruments, require immediate, real-time responses, while with other applications, like composition, real-time operation may not always be as important.

Now, let's consider these three questions as we look at some of the earliest examples of computer music.

11.2 Computer music on mainframe computers

In the 1950s, Bell Labs in New Jersey was one of the most famous research and development labs in the world. Building from the legacy of Alexander Graham Bell's invention of the telephone, this division of AT&T had created innovative new technologies that still impact how we transmit radio signals, synthesize speech, encrypt communications, and build computers.

Max Matthews was an electrical engineer and violinist working at Bell Labs. He was working with giant mainframe computers like the IBM 7094, which cost millions of dollars, occupied an entire room, and relied on punch cards and magnetic tapes to read and write data. These computers were used for many purposes, including during some of the early NASA space missions (Figure 11.1).

Figure 11.1 The IBM 7094 computer at NASA. Public domain. Available at Wikipedia.

Matthews wanted to make music with these mainframe computers. He created a programming language called MUSIC, then later MUSIC II, MUSIC III, and so on. Today we call his languages MUSIC-N (N is like a variable representing the specific version number of MUSIC).

In the MUSIC-N languages, programmers created an "orchestra" of musical instruments. The orchestra defined a set of instruments and configured how each instrument created sound. Here is an example of an instrument definition in CSound, a music programming language inspired by MUSIC-N that you can still use today:

```
        instr 2
a1      oscil         p4, p5, 1 ; p4=amp
        out           a1        ; p5=freq
        Endin
```

Even though the syntax of the programming language may look strange, the elements are actually familiar. The first line defines the instrument, which is similar to defining your own function in TunePad or EarSketch. The third line outputs the sound, which is similar to a return statement in your EarSketch or TunePad function definition. The second line calls a unit generator (oscil), which is similar to calling a function, and passes three arguments to it: p4, p5, and 1. Oscil is an oscillator that synthesizes a simple waveform, like a sine wave or square wave. The arguments configure things like how low or high the sound is (frequency) and how loud it is (amplitude).

Once a programmer has created an orchestra of instruments, they create a score with the notes to be played in a song. Here is an example of a score in CSound:

```
;ins strt dur amp(p4) freq(p5)
i2   0    1   2000    880
i2   1.5  1   4000    440
i2   3    1   8000    220
i2   4.5  1   16000   110
i2   6    1   32000   55
```

In this score, each line creates a single note of music. The line specifies which instrument will be used (in this case, the instrument 2 we just created), when the note starts and its duration (in seconds), how loud it is, and how high or low the note is (frequency in Hertz). If you want to learn more about how MUSIC-N and CSound work, you can download and run the open-source CSound software and read one of the many online CSound tutorials (https://csound.com/). These languages have been used by tens of thousands of musicians to create a wide variety of music.

With MUSIC-N, the computer took on the role of a performer. The programmer listed the notes in the musical composition (in symbolic form) and configured the instruments to play each note. Then the programmer executed the program and waited (a long time!). After the program completed, they could listen to the audio that it had generated.[1]

At the same time that Matthews and others at Bell Labs were working on MUSIC-N, researchers at the University of Illinois envisioned a different role for computers in music. Their ILLIAC I was similar in size and complexity to the IBM mainframes at Bell Labs. Unlike the Bell Labs computers, though, the ILLIAC I could not yet convert data to a format that could be played through a loudspeaker. So Lejaren Hiller and Leonard Isaacson, who were trained as both chemists and musicians, decided that instead of asking the computer to perform music, they would ask it to compose music.

In 1957, the ILLIAC I composed its first piece of music, the ILLIAC Suite.[2] Hiller and Isaacson had the computer generate random musical notes. Then the computer checked the random notes to see if they followed common musical rules of melody and harmony. If those rules were not followed, they would generate new random notes. This process continued in a loop until notes were found that satisfied all the rules. They then transcribed the program's output into traditional music notation and asked a string quartet to play it.

Many parts of their algorithm are similar to code you have written in TunePad and EarSketch. Their code would ask questions like "Do three note repeat?", which is equivalent to a conditional if-else statement you would write in Python. Their algorithm follows different branches depending on whether the answer to the question is yes or no, similar to the use of Boolean True or False values. They also incorporated subroutines, such as the "try again". A subroutine is similar to a function.

Matthews' MUSIC-N language and Hiller and Isaacson's ILLIAC Suite are two prominent examples of early computer music for large, mainframe computers. They approached the challenge of computer music in different ways. With MUSIC-N, the computer followed the instructions of a human programmer, generating each note of a song as specified in a "score" file. With the ILLIAC, the computer made random decisions and checked them against musical rules to create its own score.

There was one important similarity between the systems: they were both extremely slow. It could take hours or more to generate even a short piece of music. Today, we take for granted the instant feedback of computers and lose patience when it takes a few seconds for an app to open or a video to stream. Matthews, Hiller, and Isaacson had to reserve time on a large computer that was shared by many people and then wait patiently for their program to run. If the program had a bug, or if the music did not turn out as expected, they would make changes and then wait hours or even days before they could run their program again.

11.3 Digital synthesizers and personal computers

Early computers were large, expensive, and mostly limited to large companies and universities. Places like Bell Labs and the University of Illinois would invite a few lucky musicians to visit their labs and create new music, but most musicians had no access to computers.

In the 1980s, the popularity of two new kinds of products quickly changed this situation. First, digital synthesizers became affordable and widely available. These synthesizers combined a musical keyboard, an embedded computer for sound synthesis, and a simple interface of buttons, knobs, and sliders to create a new kind of musical instrument. Musicians could play the musical keyboard to generate a wide variety of sounds in real time.

Personal computers became small, relatively inexpensive, and easier to use. Musicians could now purchase a computer for their home or their studio. While these early personal computers had limited ability to play and record sound, a new protocol called MIDI (Musical Instrument Digital Interface) connected computers to digital synthesizers so they could make music together.

John Chowning's work in frequency modulation (FM) synthesis perfectly illustrates this transition. Chowning, a composer and professor at Stanford, invented new ways to combine oscillators together to create a variety of complex and realistic sounds. He initially discovered this technique in 1967 and published a research paper about it in 1973. However, he could only run his FM synthesis software on large computers that few people had access to. His software also took a long time to run on those computers; it could not make sound in real time. In the 1970s, Chowning composed many incredible pieces that used FM synthesis, such as Stria

(1977).[3] Few other musicians, though, were able to create their own music with FM.

Yamaha, one of the largest musical instrument manufacturers in the world, decided to license the patent for FM synthesis from Stanford for use in their products. In 1983, they released the Yamaha DX-7. This $2,000 digital synthesizer used Yamaha's special new computer chip to do FM synthesis in real time. Now any musician could buy a Yamaha DX-7, play the synth's built-in keyboard, and perform with the incredible sounds available with FM synthesis. The DX-7, which is the best-selling synthesizer of all time, quickly became part of every major recording studio and concert stage in the world. Most top hits of the 1980s used a DX-7, including songs by Kraftwerk, U2, Stevie Wonder, Phil Collins, Whitney Houston, Depeche Mode, the Beastie Boys, and Elton John (Figure 11.2).[4]

When FM synthesis migrated from university computer systems to a widely available synthesizer, it transformed the sound of a decade of pop music. It also brought financial success both to Yamaha and to Stanford. Yamaha paid Stanford more than $20 million through their license agreement.

Now that the patent for FM synthesis has expired, anyone can create their own music products with FM. Many software programs emulate the classic sounds of the DX-7 and take it in new directions. These programs are used by top music producers today. If you want to try out FM Synthesis yourself, you can try an open-source recreation of the DX-7 online at https://mmontag.github.io/dx7-synth-js/. There are also DAW (digital audio workstation) plugins and even physical recreations of the original DX-7 available. Digital synthesizers were great at creating a wide variety of sounds in real time, and they had expressive musical keyboards that made it easy to perform with them in studio or on stage. Their user interfaces, however, were difficult to use. They usually consisted of sliders, knobs, buttons, and a display that was more like the one on your microwave oven than the one on your computer. It was hard to use these synths

Figure 11.2 The Yamaha DX-7.

to design custom sounds beyond the presets, and it was hard to use them to sequence music. (Sequencing means recording, editing, and recombining sections or tracks of music to create a full song.)

Early personal computers, on the other hand, had large screens, keyboards, and sometimes mice, which made them great for displaying, entering, and editing information. They had limited ability, however, to make sound—especially in real time.

In 1983, the same year that Yamaha released the DX-7, a music industry group created MIDI, a standard that helped different music devices to communicate. When connected with MIDI cables, devices could send and receive messages to play notes, change settings, synchronize clocks, and perform other common musical tasks. MIDI remains an industry standard today.

While it was useful to connect two synthesizers together so that you could play the sounds on one with the musical keyboard on another, things got really interesting when a personal computer used MIDI to connect to digital synthesizers. Now, each device could focus on what it did best. Musicians could use their computers to sequence music and edit sounds. They could use their synthesizers to perform (on the musical keyboard) and create the actual musical sounds. The personal computer quickly became the hub of the modern music studio. MIDI acted as a musical network that connected everything together.

MIDI sequencers quickly became the most popular music application for personal computers. Sequencers handled the recording, editing, and playback of multiple tracks of MIDI data. A musician could create a song step by step, playing each track on a MIDI keyboard. They could go back and edit that MIDI data note by note, change the design of the sounds on each synthesizer, or even change the tempo. Some design elements of these MIDI sequencers, like a grid view for displaying music notes and a multi-track view of a song, remain common in music software today, including EarSketch and TunePad (Figure 11.3).

By the late 1990s and 2000s, personal computers were becoming fast enough to process audio in real time on their own, without depending on separate digital synthesizers. Over the last 20 years, the computer music studio has gradually shrunk in size. Many musicians now compose and produce music using only a laptop, or even just a tablet or phone. The MIDI sequencer has evolved into a digital audio workstation such as GarageBand, Reaper, FruityLoops, Live, Logic, and Pro Tools. A digital audio workstation can record, edit, and process audio and MIDI side by side, so musicians can choose when symbolic or audio representations work best for their project. Standalone digital synthesizers have turned into software devices and plugins that load inside of the digital audio workstation. The fundamental techniques, workflows, and designs, though, remain the same as in the early days of music software for personal computers.

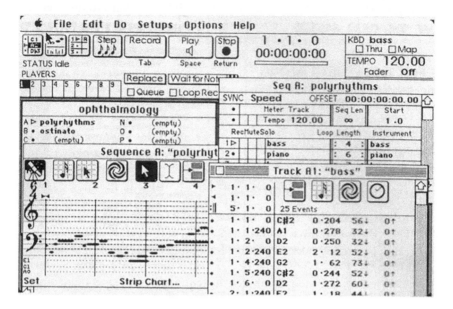

Figure 11.3 Opcode's Vision software for Macintosh in 1989.

11.4 Intelligent machine musicians

In most of the computer music applications we have considered, people tell a computer what to do. Programmers create a score and orchestra file for MUSIC-N. Musicians play a digital synthesizer to perform a song. Composers, audio engineers, and producers record, sequence, edit, and mix a song in a digital audio workstation. Even when you write code in TunePad or EarSketch, you usually specify exactly how to combine musical notes or audio clips to create a song. In all of these examples, the computer is a tool that we use to create the music we imagine.

In contrast, Hiller and Isaacson's ILLIAC Suite saw music as a collaboration between human and machine. The human creates the algorithm, but through a series of carefully controlled random decisions, the computer creates music that the human had never imagined. Both the human and the machine are—in a sense—creative. Together, they create music that neither would have made alone.

Many computer musicians have explored how computers can become creative musical collaborators rather than just musical tools. Laurie Spiegel, a composer who worked at Bell Labs, created a software program in the 1980s to address this question. Her program, Music Mouse, algorithmically generated music based on how a user moved their mouse across a grid. The user could also configure settings like scale, tempo, and harmony. In the manual for the Music Mouse software, Spiegel explained her motivation for creating it:

> Up to this time, of the new powers which computers bring to music, commercially available music software has focused mainly on precision and memory. These are wonderful attributes, but one of the computer's greatest strengths remains barely touched. Logic, the computer's ability to learn and to simulate aspects of our own human intelligence, lets the computer grow into an actively participating extension of a musical person, rather than just another tape recorder or piece of erasable paper.[5]

Music Mouse was so popular that you can still try it out today in a free web-based version (https://teropa.info/musicmouse/). Music Mouse also inspired many of the interactive music-making apps available today for mobile phones and tablets (Figure 11.4).

While Spiegel was interested in how a standard computer mouse could help musicians collaborate with computers, others preferred to use more traditional musical instruments to collaborate. In the 1980s, jazz trombonist and scholar George Lewis developed Voyager, a software program that improvised with a human musician. As the human musician plays, Voyager converts their performance into a collection of MIDI notes, it analyzes those notes, and it algorithmically generates its own music. This interaction is not a simple call and response. Lewis explains that Voyager employs a "nonhierarchical, improvisational, subject-subject model of discourse" deeply rooted in African American aesthetic traditions. Voyager controls a virtual orchestra that, unlike the orchestra in Matthew's MUSIC-N, has no pre-determined score or structure. Lewis writes: "Voyager asks us where our own creativity and intelligence might lie—not 'How do we create intelligence?' but 'How do we find it?' Ultimately, the subject of Voyager is not technology or computers at all, but musicality itself."[6] Many musicians and computer scientists are still developing new ways for computers to improvise with musicians, including Gil Weinberg, a music technology professor at Georgia Tech whose robot Shimon can play the marimba in a band (https://www.shimonrobot.com/) (Figure 11.5).

Music Mouse and Voyager both focus on collaborative music performance between human and machine musicians. Many computer musicians have also focused on music composition, going all the way back to the ILLIAC Suite. In the 1980s and 1990s, composer David Cope created the Experiments in Musical Intelligence (EMI) system. EMI analyzed musical scores of composers like Johann Sebastien Bach and Scott Joplin and then created its own scores in a similar style. More recently, advances in the field of deep learning in computer science have renewed interest in this approach. Major tech companies and startups have developed new AI systems to compose music, such as Google's Magenta

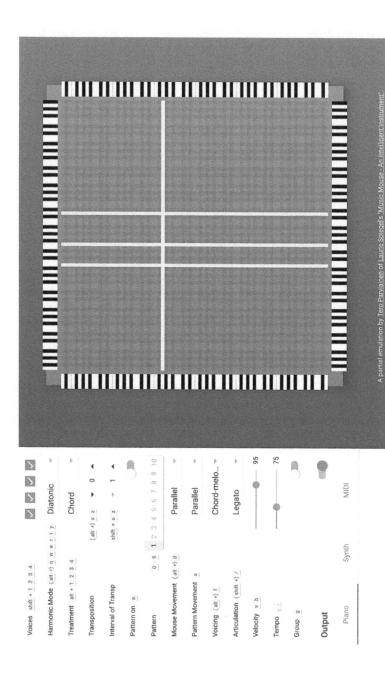

Figure 11.4 Screenshot of Teropa's Music Mouse emulator.

Figure 11.5 Jazz trombonist and scholar George Lewis working on the IRCAM project. *Image Credit: Michel Waisvisz archives, ca. 1985 (used with permission).*

(see https://magenta.tensorflow.org/). These companies see commercial potential in computers composing music, the same way that Yamaha saw commercial potential in the FM synthesis technology. They also realize that these computer composers will work best when they collaborate with human musicians instead of trying to replace human musicians. Startup Amper Music (https://www.ampermusic.com/), for example, develops software for podcast and video producers. It helps them create AI-generated music that is perfectly tailored to match the timing, style, and emotion of their content.

11.5 Conclusion

Computer music has come a long way from punch cards on mainframe computers to today's digital audio workstations and AI systems. Throughout the history of computer music, musicians, computer scientists, and engineers have collaborated to discover new ways that computers can help us to create, perform, and share music. Today, computers impact every aspect of music creation, performance, recording, production, and distribution. While early computer music was limited to those who had access to specialized equipment at universities and industry labs, today anyone with a laptop or a mobile device can make music with computers.

Learning to make music and write code with TunePad and EarSketch helps you learn many important concepts in computer music, including digital audio workstations, synthesizers, effects, and sequencing. To learn more about computer music, you can download and learn to use more computer music programming languages, such as Pure Data or SuperCollider, or download and learn to use digital audio workstation software, such as GarageBand, Reaper, or Studio One. Many middle schools and high schools now have music technology labs and teach music technology courses. Music technology (or computer music) is also a growing field of study at colleges and universities such as Georgia Tech (http://music.gatech.edu/prospective-students).

Notes

1 You can listen to an early recording demonstrating MUSIC-N here: https://www.youtube.com/watch?v=wjMaTzDh8I8.
2 You can listen to the ILLIAC Suite at https://www.youtube.com/watch?v=n0njBFLQSk8.
3 You can listen to an excerpt of Stria at https://www.youtube.com/watch?v=988jPjs1gao.
4 This playlist highlights some of the iconic songs that have used the Yamaha DX-7: https://www.youtube.com/playlist?list=PLsAmsnfaNA4-CevYq1hrcD_aIP5xWe9Xr
5 Quoted from the Music Mouse manual, which is available at http://retiary.org/ls/progs/mm_manual/mouse_manual.html. Some recent mobile apps related to Music Mouse include Thicket (http://apps.intervalstudios.com/thicket/), Borderlands Granular (http://www.borderlands-granular.com/app/), and Bloom (http://www.generativemusic.com/bloom.html).
6 George Lewis. Too Many Notes: Complexity and Culture in Voyager. Leonardo Music Journal, Volume 10, 2000.

Appendix A
Python reference

This appendix provides a quick reference for the most important Python concepts that we've covered in this book. For more on Python, check out some of the many free resources available online.

1 What is Python?

Python is a computer programming language developed in the 1990s that is now one of the most widely used languages in the world. Python is a good choice for creating things like apps, web services, video games, digital music, and art. It's also a great tool for processing large amounts of data, which makes it a popular tool for data science and machine-learning applications.

2 What does a Python program look like?

Python programs consist of lines of text written according to strict grammatical rules. The rules of a programming language are called its syntax. Python has a relatively clean and simple syntax that's easy to read and write, which is one of the reasons that it's so popular for beginners and experts alike. Here's a two-line program that runs inside of TunePad. This program prints out some text and then plays a middle C note on the piano:

```
print('This is a program')  # print text to the console
playNote(60)                 # play middle C for one beat
```

3 Comments

In the code above, some of the text appears after hashtag (#) symbols on each line. This text is called a comment—a freeform note that programmers add to make their code easier to understand. Comment text is ignored by Python, so you can write anything you want after the hashtag symbol on a line. You can also use a hashtag at the beginning of a line to temporarily disable code. This is called "commenting out" code.

4 What is a syntax error?

If you don't type your program exactly according to the rules of Python, you get an error message called a syntax error. Here's what a syntax error looks like in TunePad:

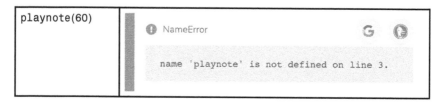

A syntax error usually includes a line number showing where the problem is. If you get a syntax error, it means that Python is confused—it doesn't understand what you're telling it to do. Because Python is confused, the syntax errors that Python generates often don't make much sense to human programmers either. In this example, Python is confused because we wrote playnote (with a lowercase "n") instead of playNote (with an uppercase "N").

Here's another syntax error where the programmer used smart quotes (" ") instead of straight up and down quotes. Unfortunately, Python gets confused when you use smart quotes.

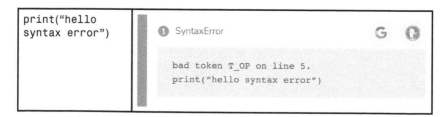

Here's one last syntax error. This one's tricky because the problem is actually on line 1 even though the syntax error says line 2. The actual problem is a missing right parenthesis on line 1.

`playNote(60` `rest(1)`	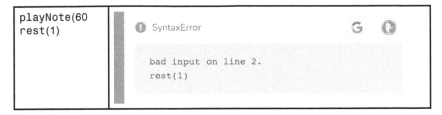

5 Functions

Almost everything you do in Python involves *calling* functions. A function (sometimes called a command or an instruction) tells Python to do something or to compute a value. For example, the **print** function tells Python to output a line of text. In the early days of computers, a print function call would literally send output to a physical printer that would type lines of text on paper tape. Modern languages do essentially the same thing only they print text to the screen instead of paper.

5.1 Calling functions

There are three parts to every function call:

1. First, you have to write the name of the function. Functions have one-word names (no spaces) that can consist of letters, numbers, and the underscore _ character. If you want a multi-word function, you typically separate the words with the underscore character as in:

   ```
   my_multi_word_function()
   ```

   or you can capitalize each new word as in playNote.
2. Second, you have to include parentheses after the name of the function.
3. Finally, you include any parameters that you want to pass to the function. A parameter provides extra information or tells the function how to behave. For example, the print statement above has one parameter, which is the text to output. This playNote function tells TunePad to play a middle C (note 60) for two beats. The first parameter is the note, and the second is the duration of the note (Appendix A1.4).

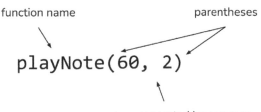

5.2 Optional parameters

Sometimes parameters are optional, meaning that they have a value that gets provided by default if you don't specify one. For playNote, only the first parameter, the note number, is required. If you don't pass a second parameter, it assumes a value of 1 beat. You can also include the names of parameters in a function call. For example, all four of these lines do the same thing; they play a note for 1 beat. The first two use parameters without their names. The second two include the names of the parameter, followed by the equals sign (=), followed by the parameter value.

```
playNote(60)                        # the beats parameter is optional
playNote(60, 1)                     # with the beats parameter set to 1
playNote(60, beats = 1)             # with a parameter name for beats
playNote(note = 60, beats = 1)   # with a parameter name for note and beats
```

5.3 Return values

Functions can also be used to compute values or produce results. When a function produces a result, it's called a return value. For example, the random function computes a random decimal number somewhere between 0.0 and 1.0. This can be particularly useful in music for creating a more human feel. In Python the random function is part of a module also called random that must be imported first.

```
from random import random   # import the random function
duration = random()         # set duration to a value between 0.0 and 1.0
playNote(60, duration + 1)  # play a note with a randomized duration
```

5.4 Defining your own functions

You can also define your own functions in Python using the **def** keyword. Once you define a function, you can use it just like any other function. For example, the code below defines two functions called **chorus** and **groove**. You can give your functions any name you want as long as they're one word long (no spaces) and consist only of letters, numbers, and the underscore character (_). Function names cannot start with a number.

```
def chorus():
    playNote(60)
    playNote(61)
    playNote(65)
```

```
def groove():
    playNote(0, beats = 0.5)
    playNote(4, beats = 0.5)
    playNote(2, beats = 0.5)
    playNote(4, beats = 0.5)
    playNote(0, beats = 0.5)
    playNote([0, 4], beats = 0.5) # double kick with hat
    playNote(2, beats = 0.5)
    playNote(4, beats = 0.5)
```

Python uses indentation to figure out which code is inside of a function and which code is outside. In the above examples, the code inside each function is indented by four spaces (Appendix A1.5).

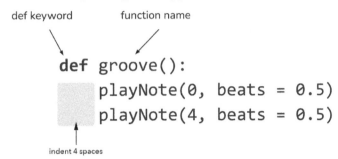

Once a function is defined, we can call it just like we would any other function:

```
chorus()
chorus()

groove()
```

5.5 *Defining functions with parameters*

You can also define functions that take parameter values. A parameter is like a special variable that you can use just inside of a function (see the next section for more on variables). Here's a quick example of a function that takes two parameters as input values. The first is the note to play, and the second is the note duration.

```
def playThreeNotes(note, duration):
    playNote(note, beats = duration)
    playNote(note, beats = duration)
    playNote(note, beats = duration)

playThreeNotes(60, 2) # play note 60 three times for 2 beats each
```

5.6 Defining functions with optional parameters

It's possible to make some of a function's parameters optional by providing a default value for them. We can change the playThreeNotes function so that the duration parameter has a predefined value.

```
def playThreeNotes(note, duration = 1):
    playNote(note, beats = duration)
    playNote(note, beats = duration)
    playNote(note, beats = duration)
```

Now when we call the function, we can decide whether or not to provide the duration:

```
# the first four are equivalent
playThreeNotes(60, 1)
playThreeNotes(60)
playThreeNotes(60, duration = 1)
playThreeNotes(note = 60, duration = 1)

# these two are not equivalent
playThreeNotes(61, 2)
playThreeNotes(61)
```

5.7 Defining functions with return values

You can define functions that produce a result using the return keyword. Functions can return any kind of value you want (see Section 7 for different data types). If a function doesn't have a return statement, it automatically produces a value called None.

```
def square(number):
    return number * number
square(6) # returns 36
square(9) # returns 81
square(-1) # returns 1
```

6 Variables

A variable is a name you give to some piece of information in a Python program. You can think of a variable as a kind of nickname or alias. For example, the code on the left plays a drum pattern without variables, and the code on the right plays the same thing with variables. Notice how the variables help make the code easier to understand because they give us descriptive names for the various drum sounds instead of just numbers.

```
1 | playNote(0)
2 | playNote(4)
3 | playNote(2)
4 | playNote(4)
```

⇨

```
1 | kick = 0
2 | hat = 4
3 | snare = 2
4 |
5 | playNote(kick)
6 | playNote(hat)
7 | playNote(snare)
8 | playNote(hat)
```

In the version on the right we defined a variable called **kick** on line 1, a variable called **hat** on line 2, and a variable called **snare** on line 3. Each variable is *initialized* to a different number value representing the corresponding drum sound. It's also possible to change the value of a variable later in the program by assigning it a different number.

```
kick = 0
playNote(kick)
kick = 1        # set kick to a different value
playNote(kick)
```

Variable names can be anything you want as long as they're one word long (no spaces) and consist of only letters, numbers, and the underscore character (_). Variable names cannot start with a number, and they can't be the same as any existing Python keyword.

7 Data types

Variables can hold different kinds of data. There are five common data types in Python:

```
name = "C major" # text values (string)
coda = False      # a value that is either True or False (boolean)
outro = None      # a value that means nothing or no value
note = 17         # integer number (int)
duration = 1.75   # decimal number (float)
```

We can change the value of a variable as a program runs. We can also change the data type of a variable when we assign it to a new value. Python decides what type a variable is by the way it's declared. This can sometimes be confusing for numeric data types because some of the math operators behave differently if a variable is an integer or a decimal (floating point) number.

7.1 Strings

You can think of a *string* as text or a list of characters. For example: "**Hello, I'm a string**" has 18 characters starting with "**H**" and ending with "**g**" and including the spaces. Python uses either double quotes (") or single quotes (') to define a string (remember that smart quotes won't work).

```
# you can create a string with either double or single quotes
first_name = 'Ella'
last_name = "Fitzgerald"
```

You can access individual characters in a string using index notation. For example, these two lines of code get individual letters out of the **last_name** variable. Notice that Python uses zero as the index of the first character in a string.

```
letter = last_name[0]   # letter is equal to 'F'
letter = last_name[3]   # letter is equal to 'z'
```

7.2 Booleans

The second data type is called a **boolean**. A boolean is a value that can only be **True** or **False**. Booleans are useful for asking yes or no questions in code, which is a common thing to do in programming.

7.3 Floats

Floating point numbers (or floats) are real numbers with a decimal point that can be positive, negative, or zero.

7.4 Ints

Integers (or ints) are whole numbers without a decimal point that can be positive, negative, or zero.

7.5 None

The **None** value is a way of saying that a variable has no current value, which can also be useful in programming from time to time.

8 Lists

Variables are great for working with individual pieces of data, but often we want to work many data elements at the same time. One of the reasons programming languages are powerful is that they can operate on large arrays of data stored in the computer's memory. In Python there's a data type called a *list* that lets us keep track of an arbitrary number of data elements. We can create a list using square brackets with comma separated data elements inside. Lists are very useful in TunePad for playing chords (more than one note at the same time). For example, this line of code uses a list to play a C major chord.

```
playNote([ 60, 64, 67 ], beats = 4)
```

The notes of the chord are enclosed in square brackets **[60, 64, 67]** to create a list. Just as with other data types that we've introduced so far, you can use a variable to refer to a list.

```
Cmajor = [ 60, 64, 67 ]
Dminor = [ 62, 65, 69 ]
playNote(Dminor)
playNote(Cmajor, beats = 4)
```

This code creates two variables called Cmajor and Dminor that each hold a list of numbers representing the notes of the chord. Python lists are versatile; they come with all sorts of useful features, the most important of which we describe below.

8.1 Accessing individual list elements

You'll often want to be able to access individual elements of a list once it's defined. In Python, the first element of a list is stored at index 0, and you can access it like this:

```
first_note = Cmajor[0] # get the first note of the Cmajor chord
playNote(first_note)
```

The second element is at index 1, the third at index 2, and so on. The fact that the first element of a list is stored at index 0 takes some getting used to and is a common source of confusion for beginners.

To access elements from the end of a list, Python uses negative numbers. The last element of a list is stored at index -1, the second-to-last element is stored at index -2, and so on.

```
last_note = Cmajor[-1] # get the last note of the Cmajor chord
playNote(last_note)
```

8.2 Changing list elements

Indexing can also be used to modify a list element. If we wanted to change the first note of the Cmajor chord we could use this syntax.

```
Cmajor[0] = 48 # change the first note of the scale
```

8.3 Length of a list

The **len()** function tells you how long a list is.

```
print(len(Dminor)) # outputs the number of notes in the chord
```

8.4 Adding to a list

Python lists can grow and shrink. To add an element to the end of a list, use the **append** function. This function is *part* of a list, and uses a slightly different kind of syntax. We have to put the list followed by a period (dot), followed by the **append** function name. The **append** function takes one parameter, which is the value to add to the end of the list.

```
Dminor.append(72)    # add a C note to the end of the Dminor chord
print(Dminor)        # prints [ 62, 65, 69, 72 ]
print(len(Dminor))   # prints 4
```

We can also insert elements into arbitrary positions in a list using the **insert** function. To change the example, let's try a list of words instead of numbers.

```
words = [ "cat", "dog", "mouse", "horse", "goat" ]
words.insert(1, "elephant") # elephant will now be the 2nd element
```

To insert an element at the very beginning of the list we would use **insert** with index 0.

```
words.insert(0, "moose") # moose will now be the 1st element
```

8.5 Removing from a list

There are several ways to remove elements from a list. One of the most direct is to call the **remove** function on a list.

```
words = [ "cat", "dog", "mouse", "horse", "goat" ]
words.remove("dog") # the second element will now be removed.
```

A second option is to call the **pop** function to remove and return the last element of the list.

```
words.pop() # returns "goat" which will now be removed from the list.
```

The **pop** function can also be called with an index that says which element to remove.

```
words = [ "cat", "dog", "mouse", "horse", "goat" ]
```

```
words.pop(2) # removes and returns "mouse"
```

You can remove all of the elements of a list using the **clear** function.

```
words = [ "cat", "dog", "mouse", "horse", "goat" ]
words.clear() # the list is now empty
```

8.6 Selecting sub-lists

Python provides powerful tools for accessing segments or *slices* of larger lists. In practice, this looks similar to list indexing, but allows us to return a list of elements rather than a single element. We do this by specifying start and stop indices, separated by a colon:

```
myList[start:stop]
```

We can use the result of our slice operation as we would any other list:

```
scale = [60, 62, 64, 65, 67, 69, 71, 72]
print(scale[2:4]) # prints [64, 65]
for note in scale[0:5]:
    playNote(note)
```

Neither of these parameters is required for slicing. If we omit the **stop** parameter, then our slice goes through the end of the list. If we omit the **start** parameter, the slice begins at the beginning of the list. Since slicing returns a copy of the specified elements, if we wanted to copy the entire list, we would omit both start and stop parameters:

```
myList[:]
```

Syntax	Example	Explanation
myList[start:stop]	scale[2:4]	Returns a list of elements start through stop-1
myList[start:]	scale[2:]	Returns a list of elements from start through the end of the list
myList[:stop]	scale[:5]	Returns a list of elements from 0 through the stop-1
myList[:]	scale[:]	Returns a copy of complete list

8.7 Shuffle

You can randomly shuffle a Python list like a deck of cards using the **shuffle** function from the **random** module. This can be useful for things like playing random arpeggios.

```
from random import shuffle

notes = [ 60, 62, 64, 65, 67, 69, 71 ] # C major scale
shuffle(notes) # randomly reorders the scale
```

8.8 Choice

You can select one random element from a list using the **choice** function from the **random** module.

```
from random import choice

cards = [ 0, 1, 2, 3, 4, 5, 6, 7, 8 ]
my_card = choice(cards) # pick one number from the list at random
```

8.9 Reverse

You can reverse the order of a **list** using the reverse function of a list object.

```
notes = [ 60, 62, 64, 65, 67, 69, 71 ] # C major scale
notes.reverse() # reverse the order of the scale

for note in notes: # play notes from high pitch down to low pitch
    playNote(note)
```

8.10 Joining lists

You can join two lists together to create a longer list using the plus (+) sign.

```
Cmaj = [ 60, 64, 67 ]
Dmin = [ 61, 64, 68 ]
combined = Cmaj + Dmin # results in [ 60, 64, 67, 61, 64, 68 ]
```

8.11 Sorting lists

You can sort a list in alphabetical or numerical order using the **sort** function.

```
words = [ "cat", "dog", "mouse", "horse", "goat" ]
words.sort() # will now be in the order: cat, dog, goat, horse, mouse
numbers = [4, 3, 5, 2, 9]
numbers.sort() # will now be in the order [2, 3, 4, 5, 9]
```

9 Dictionaries

In Python, **dictionaries** or **maps** are unordered sets of data consisting of values referenced by *keys*. These keys aren't the same as musical keys. They're more like the kind of keys that open locked doors. Each different key opens its own door.

Dictionaries are extremely useful in programming because they provide an easy way to store multiple data elements by name. For example, if we wanted to store information for a music streaming service, we might need to save the song name, artist, release date, genre, record label, song length, and album artwork. A dictionary gives you an easy method for storing all of these elements in a single data object.

```
track_info = {
 "artist" : "Herbie Hancock",
 "album" : "Head Hunters",
 "label" : "Columbia Records",
 "genre" : "Jazz-Funk",
 "year" : 1973,
 "track" : "Chameleon",
 "length" : 15.75 }
```

Dictionaries are defined using curly braces with keys and values are separated by a colon. Different entries are separated by commas. After defining a dictionary, we can change existing values or add new values using the associated keys. Similar to the way we access values in a list with an index, we use square brackets and a key to access elements in a dictionary.

```
release_date = track_info['year']
track_info['year'] = 1974 # update the dictionary with a new value
```

We can also use the same syntax to add a value to a dictionary.

```
track_info["artwork"] = "https://images.ssl-images-amz.com/images/81KRhL.jpg"
```

Because the key "artwork" hasn't been used in the dictionary yet, it creates a new key-value pair. If "artwork" had been added already, it would change the existing value. One thing to notice is that values in a dictionary can be any data type including strings, numbers, lists, or even other dictionaries. Dictionary keys can be strings or numerical values, but they must be unique for each value stored.

10 The range function

The range function is used to generate an ordered sequence of numbers. This can be useful for many things in coding, especially when we want to repeat a set of actions multiple times in a row. There are three common variations of range:

```
Variation 1:          range(count)
Variation 2:          range(start, stop)
Variation 3:          range(start, stop, step)
```

In the first version, **range** takes a single parameter that tells it how long the sequence of numbers should be. For example, **range(5)** generates an ordered set of numbers that is five elements long, starting at 0 and going up to (but not including) 5:

$$0, \ 1, \ 2, \ 3, \ 4$$

In the second version, **range** takes two parameters, a starting value and a stopping value. The resulting sequence will start with the first parameter and go up to (but not include) the second parameter.

```
range(0, 7)        → 0, 1, 2, 3, 4, 5, 6
range(100, 105) → 100, 101, 102, 103, 104
range(50, 40)   → EMPTY
range(50, 50)   → EMPTY
range(-5, 0)    → -5, -4, -3, -2, -1
```

If the second parameter is less than or equal to the first parameter, the sequence will contain no values.

The final version of **range** takes a third parameter that specifies a step count (or skip count) for the sequence. For example, if you want to count by twos from 0 up to 10, you could use this statement:

```
range(0, 10, 2) → 0, 2, 4, 6, 8
```

Using the third form, you can also count backward by using a negative step count.

```
range(10, 0, -1) → 10, 9, 8, 7, 6, 5, 4, 3, 2, 1
```

11 For loops

One of the reasons we program computers is because they can do repetitive and mind-numbing tasks quickly and without error, and the ability to repeat (or loop) is a core part of almost all programming languages. There are a few different kinds of loops in Python, but one of the most common is called a for loop. Here's the basic structure of a for loop. The underlined text gets replaced with actual code that we can see in examples lower down:

```
for loop_variable in list:
    ...do something...
```

Here's a concrete example:

```
notes = [ 60, 64, 67 ] # C major chord
for n in notes:
    playNote(n)
```

The loop starts with the **for** keyword, which tells Python we're starting a loop. The next part is called the *loop variable*. This is a special kind of variable that Python will assign to each element of a list, one at a time. We can use any valid variable name for the loop variable. Next comes the in keyword followed by the list itself. This list can be a variable name, a function or expression that generates a list, or a list defined inline.

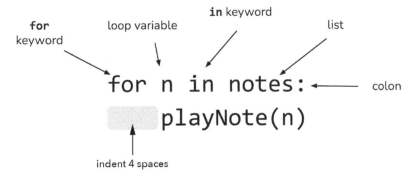

To make this clearer, each of these three loops below does the exact same thing (plays the notes of a C minor chord, one at a time):

```
notes = [ 60, 63, 67 ]          # C minor chord

for n in notes:                 # the list is a variable
    playNote(n)

for n in [ 60, 63, 67 ]:        # the list is defined inline
    playNote(n)

for n in majorChord(60):        # a function generates the list
    playNote(n)
```

The next part of the loop syntax is the colon character that tells Python that we're going to start a *block* of code that will get run each time the loop repeats. In the examples above, we only did one thing inside the loop, but you can include as many lines of code as you want. For example, this loop prints out a list of names and then prints out the total character length for all of the names combined.

```
1   total = 0
2   names = [ 'Herbie', 'Miles', 'Ella' ]
3   for name in names:
4       print(name)
5       total = total + len(name)
6   print(total)
```

Python uses indentation to figure out which code is inside the loop and which code is outside. In the above example, lines 4 and 5 are each indented by four spaces, which makes them part of the code that gets repeated in the loop. Line 6 is not indented, which means that it only gets run one time after the loop is finished repeating. If you run this code, the output is:

```
Herbie
Miles
Ella
15
```

Note that the list we use in a for loop can contain any data type, not just numbers.

11.1 *For loops with the range function*

One of the most common ways to use a for loop is in combination with the range function to repeat an action for a fixed number of times. When we're creating music with code, this kind of loop is useful for things like high-hats runs and other repeated patterns (Appendix A1.7):

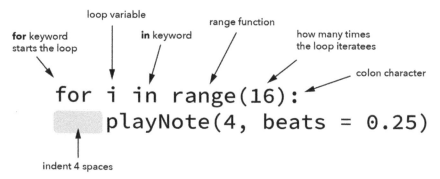

A for loop with the range function:

- begins with the keyword **for**
- includes a loop **variable** name; this can be anything you want (above it is "i")
- includes the keyword **in**
- includes the **range** function
- says how many times to repeat (above, this is 16)
- includes a colon **:**
- includes a block of code indented at four spaces

The indented block of code is repeated the total number of times specified by **range**. Each time the loop goes around, the variable is incremented by one.

The loop variable can also come in handy here if we want to do something like creating a crescendo where each note gets successively louder.

```
for vol in range(10, 101, 10):
    playNote(50, velocity = vol)
```

...or a decrescendo where the volume gets softer and softer:

```
for vol in range(100, 0, -10):
    playNote(58, velocity = vol)
```

11.2 Nested loops

Loops can be combined with other programming constructs in complex ways. One common tool is to *nest* one loop inside another. For example, suppose we wanted to generate a snare drum riser that plays four sets of eight notes with each new round of notes played twice as fast as the previous round. This creates an accelerating pattern that looks something like this (Appendix A1.8).

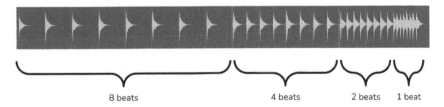

One way to do this is to use a nested for loop.

```
1  duration = 1.0                # duration of each note
2  for i in range(4):            # 4 sets of notes
3      for j in range(8):        # each set has 8 notes each
4          playNote(2, beats = duration)
5      duration = duration / 2   # cut the note duration in half
```

The outer loop repeats four times (line 2), and the inner loop repeats eight times (line 3). This means that the **playNote** on line 4 gets called 8 × 4 = 32 times. The length of each note is determined by the **duration** variable that we declare on line 1. This variable starts out with a value of 1.0, but in each iteration of the outer loop, we divide the value in half.

12 Conditional logic

Conditional logic is a way of asking a question in a program based on a boolean value. If the value is True, Python will run one block of code. If the value is False, it will run a different block of code.

12.1 *If statements*

The most basic form of conditional logic is an **if statement**. The if statement runs a block of code if a value is **True** and skips it otherwise:

```
verbose = True
if verbose:
    print("I have a lot to say")
```

This example always prints out "I have a lot to say". If the **verbose** variable had been **False**, it wouldn't have printed out anything. If statements work with arbitrarily long blocks of code:

```
# assume the drum_fill variable has already been defined
if drum_fill:
    for i in range(4):
        playNote(2, beats = 0.25)
```

In this case, if the **drum_fill** variable is **True**, Python runs a for loop to play four snare drum sounds. In an if statement, the input value can be any expression or function that evaluates to a boolean. Here's an example that uses comparison and logical operators (see Section 13 below).

```
if (getPlayhead() < 2) or (getPlayhead() > 3):
    playNote(11)
```

If statements can also be nested inside of one another to create more complex conditional logic.

```
if testA:
    if testB:
        # run some code if both testA and testB are both true
        print('A and B are both True')

    # run some code if testA is true
    print('A is True. B may or may not be True')
```

12.2 If else statements

A variation of the if statement allows us to run alternative code if our boolean value evaluates to **False**. Here's a musical example that either plays some notes *or* rests depending on the situation.

```
#... assume decoration is already defined
if decoration:
    playNote(24)
    playNote(27)
else:
    rest(2)
```

In this example, if **decoration** is **True**, the two **playNote** functions get called. Otherwise, if decoration is **False** the **rest** statement gets called.

12.3 If Elif statements

In a third variant of the if statement let's chain an arbitrary number of alternative conditions together. This example uses a logical operator called == to test if two values are the same.

```
volume = "medium"
if volume == "loud":
    playNote(60, velocity = 100)
elif volume == "medium":
    playNote(60, velocity = 80)
elif volume == "soft":
    playNote(60, velocity = 30)
else:
    playNote(60, velocity = 0)
```

In this code the **elif** stands for "else if" and allows us to ask a series of questions. Whichever **if** or **elif** statement evaluates to **True** first is run; the other statements are ignored. If nothing is **True**, then the **else** condition is run if it is provided. These examples are not very exciting yet, but things get more interesting when we combine if statements with logical operators.

13 Python operators

Python uses *operators* to compare or transform values. Some of the most common operators test for equality and inequality or perform arithmetic operations such as addition, subtraction, and multiplication. Operators can be combined into compound statements often using parentheses.

13.1 Comparison operators

Comparison operators compare two numeric values and return True or False depending on the result.

Code	Name	Description	Examples
==	Equal	Returns True if two values are the same; returns False otherwise.	a == b
!=	Not Equal	True if two values are not the same.	a != b
>	Greater Than	True if the first number is greater than the second.	a > b
<	Less Than	True if the first number is less than the second.	a < b
>=	Greater Than or Equal To	True if the first number is greater than or equal to the second.	a >= b
<=	Less Than or Equal To	True if the first number is less than or equal to the second.	a <= b

13.2 Logical operators

Logical operators combine boolean values together into compound logical statements.

Code	Description	Examples
and	Returns True when both values are True; returns False otherwise.	a and b (x > 0) and (x < 10)
or	Returns True if either value is True; returns False if both values are False.	a or b (note == 60) or (note == 62)
not	Returns True if the value is False; returns False if the value is True.	not a not (pop or rock)

13.3 Containment operators

Containment operators allow you to test whether or not a value is contained in a list or sequence.

Code	Description	Examples
`in`	Returns True if a value is contained in a list.	`notes = [ 60, 64, 67 ]` `65 in notes`
`not in`	Returns True if a value is not contained in a list.	`65 not in notes`

13.4 Arithmetic operators

Arithmetic operators perform common mathematical operations with numbers.

Code	Name	Description	Example
`+`	Addition	Adds two numbers	`a + b`
`-`	Subtraction	Subtracts the second number from the first	`a - b`
`*`	Multiplication	Multiplies two numbers together	`a * b`
`/`	Division	Divides the first number by the second. Division always returns a decimal number	`a / b`
`//`	Integer Division or Floor Division	Divides the first number by the second, rounding down to the nearest integer. The result will be an integer if both numbers are integers. Otherwise the result is a decimal	`a // b`
`**`	Exponent	Raise the first number to the power of the second. The example shows x^3	`x ** 3`
`%`	Modulus or remainder	Performs a division and returns the remainder. For example, 3.5% 2 would be 1.5. Or 5% 3 would be 2	`3.5% 2` `5% 3`

13.5 Order of operations

Python has precise rules for how it evaluates expressions using operators. Some of these rules are familiar from math class—multiplication and division take precedence over addition and subtraction, To avoid confusion, it's good practice to enclose different operators in parentheses.

```
x ** 3 > 15 or y < 8.3 / x + 0.5 # CONFUSING!!
```

```
((x ** 3) > 15) or (y < (8.3 / (x + 0.5))) # Less confusing
```

14 While loops

A while loop is a way of repeating a set of instructions *until* some condition is met. This is useful when we don't know in advance how many times we'll need to loop. Here's an example that loops until a sound becomes too quiet to hear:

```
volume = 100
while volume > 20:
    playNote(10, beats = 0.5, velocity = volume)
    volume = volume * 0.9 # reduce to 90% of volume
```

15 Objects and classes

Nearly everything in Python is an **object**: lists, variables, and even functions. Objects are abstractions for data. They describe how the data is represented and manipulated. Python also gives programmers the flexibility to define their own objects, called classes. Classes give programmers the ability to precisely structure the data they are working with and to define functions, called **methods**, for working with that data. The individual pieces of data that belong to a class are called fields. These fields can be made up of anything: numbers, strings, lists, dictionaries, and even instances of other classes.

Imagine we want to define a class to describe any person. This class should have two fields: name and age. We can specify initial values of these fields by defining a special method called **__init__**. This method will take the parameter **self** and whichever fields we want to specify. The **self** parameter specifies that the method is acting on the object. We can also define any other fields we want the class to have.

```
class Person:
    def __init__(self, name, age):
        self.Name = name
        self.Age = age
        self.Species = "human"
```

Let's take a quick dive into this code. On line 1, the **class** keyword tells Python that we're defining a new class, which we have called "Person". On line 2, we begin defining our initializing function which is invoked when we want to create a new instance of our object. This function takes self as well as fields for the name and age of our person. On lines 3 and 4, we create the actual fields for the object and set those fields equal to the values passed in by a user. When we want to use these values inside for these fields inside the class definition, say, in a method, we use self.*fieldName*. If we want to access them outside of the definition, we can use *variable.fieldName*. After defining our class, we can then create an instance of this class by calling it much like a function:

```
# create instance of Person and set equal to variable
p1 = Person("Miles Davis", 42)
print(p1.Age) # outputs: "42"
```

This example isn't particularly useful yet, but we can add a method to the class to work with the data. Let's define the method *ageInMonths* which returns the person's age in months. Like our __init__ function, this will also take the parameter **self**. Since our object already has an **age** field, our method doesn't need to accept any other parameters. The rest of the method declaration will look much like a typical function definition.

```
class Person:
    def __init__(self, name, age):
        self.Name = name
        self.Age = age
        self.Species = "human"

    def ageInMonths(self):
        return self.Age * 12

# create instance of Person and set equal to variable
p1 = Person("Stevie Wonder", 42)
print(p1.ageInMonths()) # outputs: 504
```

The concept of classes is a powerful tool. The examples above are somewhat trivial so now consider a class that encapsulates the idea of musical keys. How might you use it? What fields might this object have? How would you encapsulate the functions we defined in Chapters 5 and 6 to work with scales and chords?

16 Modules

One of the nice things about using Python—or any modern, fully featured language—is that it provides you with a wide variety of existing code libraries that help you with common programming tasks. Often these libraries are grouped into collections called modules. Python has a large set of modules called the *standard library*, but it also lets you develop your own modules, which is useful as you build up your toolkit as a coder. There are a number of ways to import module code into your own projects.

First, you can import an entire module into your code, which allows you to use functions, classes, and variables from that module.

```
import random
```

If we would like to use a specific item from this module, we have to refer to its namespace. The namespace gives us a clean way to track where a piece of code comes from. This becomes important for larger pieces of software. Referring to namespaces reduces ambiguity especially if different modules have functions or classes with the same name! In the example below, **random** is the namespace (i.e., the module we've imported) and *randint* is the function we are calling:

```
x = random.randint(0, 10)
```

We can also import specific items from a module. If we do these, we do not have to use the namespace.

```
from random import randint
x = randint(0, 10)
```

We can also define a new namespace using the as keyword:

```
import random as rand
x = rand.randint(0, 1)
```

Finally, if we do not want to use a namespace for the code we're importing, we can import all of the code from the module using the * symbol:

```
from random import *
```

   In TunePad, each instrument is like a module. Functions that you define in one instrument or definition cell can be imported into another if their names are valid Python variables. This becomes a powerful tool for crafting songs with repeating elements.

Appendix B

TunePad programming reference

This appendix provides a reference for the full set of TunePad functions.

---

playNote(note, beats = 1.0, velocity = 100, sustain = 0)

Play a note with a pitch value greater than 0. You can also call **playNote** with a list of notes that will be played at the same time. The optional parameter **beats** sets how long the note will last, and the optional parameter **velocity** sets how loud the note will sound. Velocity can be any number between 0 and 100. The optional **sustain** parameter allows a note to ring out longer than the value given by the beats parameter. The value of the parameter is the number of beats the note should sustain in beats.

```
playNote(32)
playNote(55, beats = 0.5, velocity = 20)
playNote(0, beats = 2, velocity = 80)
playNote([36, 40, 43])
playNote([36, 40, 43], beats = 2)
playNote(48, beats = 0, sustain = 4)
```

---

playSound(sound, beats = 1.0, pitch = 0, velocity = 100, sustain = 0)

Play a custom sound using its ID number. You can also call **playSound** with a list of sounds that will be played at the same time. The optional parameter beats sets how long the note will last, and the optional parameter **velocity** sets how loud the note sounds. Velocity can be any number between 0 and 100. The optional **pitch** parameter changes the pitch of the sound by the given number of semitones. For example, a pitch value of 3.0 would be the same as the difference from a C to a D♯ on the piano keyboard. The optional **sustain** parameter allows a note to ring out longer than the value given by the beats parameter. The value of the parameter is the number of beats the note should sustain in beats.

```
playSound(1203)
playSound(1203, beats = 0.5, velocity = 20)
playSound(1203, beats = 2, pitch = 3, velocity = 80)
playSound([1203, 559, 43])
```

---

rest(beats = 1)

Add a pause between notes. The length of the pause can be set using the optional **beats** parameter.

```
rest()
rest(2)
rest(beats = 1.5)
```

---

moveTo(time)

Moves the playhead forward or backward in time to an arbitrary position. In TunePad, the playhead marks the current time. For example, after calling **playNote(32, beats = 2)**, the playhead will have advanced by two beats. The **time** parameter is given in elapsed beats and specifies the point that the playhead will be placed. For example, a value of 0 would move the playhead to the beginning of a track (zero elapsed beats). A value of 2 would move the playhead to the beginning of the third beat (two elapsed beats).

```
moveTo(0)
moveTo(4)
moveTo(0.5)
```

---

rewind(beats)

Moves the playhead backward in time by an amount *relative to its current position*. This can be a useful way to play multiple notes at the same time.

The **beats** parameter specifies the number of beats to move the playhead. Negative values of beats move the playhead forward.

```
rewind(1)
rewind(-2)
rewind(0.5)
```

---

fastForward(beats)

Moves the playhead forward in time by an amount *relative to its current position*. This can be a useful way to play multiple notes at the same time. The **beats** parameter specifies the number of beats to move the playhead. Negative values of beats move the playhead backward.

```
fastForward(1.5)
fastForward(-2)
```

---

getPlayhead()

Returns the current value of the playhead as a float. The value returned is an elapsed number of beats. Therefore if the playhead is at the very beginning, the result will be zero rather than one.

```
getPlayhead() # returns 0.0
playNote(0, beats = 1)
getPlayhead() # returns 1.0
```

---

getMeasure()

Returns the current measure as an integer value. Note that **getMeasure** returns an elapsed number of measures. So, if the playhead is at the beginning of the track or anywhere before the end of the first measure, the function will return 0. If the playhead is greater than or equal to the start of the second measure, **getMeasure** will return 1, and so on.

```
getMeasure() # returns 0
playNote(0, beats = 4)
getMeasure() # returns 1
```

---

getBeat()

Returns an elapsed number of beats *within the current measure* as a decimal number. For example, if the playhead has advanced by a quarter beat within a measure, **getBeat** will return 0.25. The value returned by **getBeat** will always be less than the total number of beats in a measure.

```
getBeat() # returns 0

playNote(0, beats = 2)
getBeat() # returns 2

playNote(0, beats = 1)
getBeat() # returns 3

playNote(0, beats = 1)
getMeasure() # returns 0
```

---

with bend(cents = 0, beats = -1, start = 0):

Adds a pitch bend effect that changes the value of notes over time. The **cents** parameter represents the total change in pitch. One cent is equal to 1/100 of a semitone (the distance between two adjacent notes). Using a value of 500 for the cents parameter would bend the note by five semitones (the same as the distance from a C to an F on the piano keyboard). The **beats** parameter specifies how long it takes for the note to bend. If you don't provide the beats parameter, the effect will be constant. An optional **start** parameter specifies how long to wait (in beats) before starting the effect. You can also provide a list of values instead of a single number for the cents parameter. These values represent the change in cents over time. Each number will be evenly distributed over the duration of the effect.

```
# apply a constant pitch bend
with bend(cents = 100):
    playNote(36)

# bend a note from 48 to 53 over one beat
with bend(cents = 500, beats = 1):
    playNote(48)

# bouncy spring effect
with bend(cents = [0, 500, -500, 500, -500, 500, 0], beats = 1):
    playNote(42, beats = 2)
```

---

with gain(value = 0, beats = 1, start = 0):

Changes the volume of notes. The **value** parameter represents the change in volume (e.g., a value of 0.5 would reduce the volume by about half). If a single number is provided, a constant change in volume will be applied—there will be no change over time. You can also pass a list of numbers to create a change over time. Each number will be evenly distributed over the duration of the effect given by the beats parameter. An optional **start** parameter specifies how long to wait (in beats) before starting the effect.

```
# cut the volume by half
with gain(value = 0.5):
    playNote(48)
```

```
# fade in
with gain(value = [ 0, 1 ], beats = 2):
    playNote(48, beats = 2)

# fade out after one beat
with gain(value = [ 1, 0 ], beats = 2, start = 1)
    playNote(42, beats = 3)
```

---

with pan(value = 0, beats = 1, start = 0):

Applies a stereo pan effect, shifting the sound more toward the left or right speaker. The **value** parameter ranges from -1.0 (full left speaker) to 1.0 (full right speaker). A value of 0.0 evenly splits the sound. If a single number is provided, the effect will be constant—there will be no change over time. You can also pass a list of numbers to create a change over time. Each number will be evenly distributed over the duration of the effect given by the **beats** parameter. An optional **start** parameter specifies how long to wait (in beats) before starting the effect.

```
# slowly pan from the left speaker to the right over three beats
with pan(value = [-1.0, 1.0], beats = 3):
    playNote(35, beats = 3)
```

---

with lowpass(frequency = 1000, beats = 1, start = 0):

Applies a lowpass filter effect that reduces the energy of high-frequency sounds while leaving low-frequency sounds below a cutoff point unaffected. The **frequency** parameter specifies the cutoff frequency for the effect (between 10 Hz and 22 kHz). You can also pass a list of numbers to create a change over time. Each number will be evenly distributed over the duration of the effect given by the **beats** parameter. An optional **start** parameter specifies how long to wait (in beats) before starting the effect.

```
# creates a wha-wha effect after one beat by quickly changing
# the frequency cutoff of a lowpass filter between 200 and 800 Hz
with lowpass(frequency = [200, 800, 200, 800, 200, 800], beats=1):
    playNote(47, beats=3)

# adds a rhythmic pulse to piano notes
for i in range(0, 4):
    with lowpass(frequency = [ 50, 800, 50 ], beats = 0.25, start = 1):
        playNote(33, 2)
```

---

with highpass(frequency = 1000, beats = 1, start = 0):

The highpass filter reduces the energy of low-frequency sounds while allowing frequencies above the cutoff point to pass through unaltered. The

frequency parameter specifies the cutoff frequency for the effect (between 10 Hz and 22 kHz). You can also pass a list of numbers to create a change over time. Each number will be evenly distributed over the duration of the effect specified by the **beats** parameter. An optional **start** parameter specifies how long to wait (in beats) before starting the effect.

---

with bandpass(frequency = 1000, beats = 1, start = 0):

The bandpass filter allows frequencies near the cutoff point to pass through unaltered while reducing the energy of frequencies above and below. The **frequency** parameter specifies the cutoff frequency for the effect (between 10 Hz and 22 kHz). You can also pass a list of numbers to create a change over time. Each number will be evenly distributed over the duration of the effect specified by the **beats** parameter. An optional **start** parameter specifies how long to wait (in beats) before starting the effect.

```
# carving out frequencies for a clap sound in the drums
clap = 10
with bandpass(frequency = [100, 11000], beats=4):
    for i in range(0, 16):
        playNote(10, beats = 0.25)
```

---

with notch(frequency = 1000, beats = 1, start = 0):

The notch filter reduces the energy of sounds near the cutoff frequency, while allowing higher and lower frequencies to pass through unaltered. The **frequency** parameter specifies the cutoff frequency for the effect (between 10 Hz and 22 kHz). You can also pass a list of numbers to create a change over time. Each number will be evenly distributed over the duration of the effect specified by the **beats** parameter. An optional **start** parameter specifies how long to wait (in beats) before starting the effect.

---

with swing(value, beats = None, start = 0, rate= 0.5):

The swing effect introduces delays in rhythmic patterns, which adds a bounce to the rhythm's timing. This can make the performance sound more human by removing the mechanical precision. The **value** parameter is a number between 0.0 and 1.0 which specifies how pronounced this effect should be. The rate parameter specifies which note value should be swung.

```
# partly swing a 8th note snare pattern
snare = 2
```

```
with swing(0.5):
    for i in range(0, 16):
        playNote(snare, beats = 0.5)

# fully swing a 16th note hi-hat pattern
hihat = 4
with swing(1.0, rate=0.25):
    for i in range(0, 16):
        playNote(hihat, beats = 0.25)
```

Tunepad chords module

Import this module into your code using

```
from tunepad.chords import *
```

---

changeKey(key)

Specify the key of a cell with the **changeKey** function, which changes the global key of a cell. This function accepts a string from the set of possible valid keys, with the default key being "C" major.

```
changeKey("Cm")
changeKey("Bb")
changeKey("F#m")
```

---

buildChord(chord, octave=3, inversion=0)

The **buildChord** function returns a list of integers of an inputted diatonic or chromatic chord belonging to the global key corresponding to the notes specified by a Roman numeral string and inputted octave. The **octave** can be specified using an integer value in the range [-1, 9]. The **inversion** parameter controls the ordering of the pitches and has a range of [0, 3].

```
buildChord("I")
buildChord("ii", octave=[2, 3])
buildChord("V7", octave=3, inversion=2)
```

**playChord(chord, beats=1, velocity=90, sustain=0,
playType='block', octave=3, inversion=0)**

The **playChord** function plays a specified diatonic or chromatic chord for a given key. The chord is specified as a Roman numeral string. For more information on the **beats, velocity**, and **sustain** parameters, see **playNote**. The **octave** parameter specifies the octave in which the chord is played. This octave parameter is an integer in the range [-1, 9]. The **inversion** parameter controls the ordering of the pitches in the chord and has a range of [0, 3]. Using the **playType** argument allows us to choose from one of five methods of playing our chord:

- The default method is "block" and plays every note at the same time
- The "rolled" method introduces a slight, random offset between each note to mimic how a human might play the notes
- The "arpeggio" plays the chord up then down, each note an equal division of the total beats
- The "arpeggio_up" plays the chord up, each note an equal division of the total beats
- The "arpeggio_down" plays the chord down, each note an equal division of the total beats

```
playChord("I", beats=4)
playChord("ii", beats=4, octave=[2, 3])
playChord("V7", beats=4, octave=3, inversion=2)
playChord("vi", beats=4, octave=3, inversion=2, playType="arpeggio")
```

---

buildScale(tonic, octave, mode, direction="ascending")

The **buildScale** function returns a list of integers corresponding to the notes of an inputted musical scale starting on an inputted note and octave. The starting note is specified with the **tonic** parameter, a string from the set of possible valid keys. The **octave** can be specified using an integer value in the range [-1, 9]. The **direction** parameter supports "ascending" and "descending" and dictates whether the scale is in increasing order of pitch or decreasing order of pitch. The mode parameter accepts a string equivalent to:

- "Major" or "minor"
- One of the church modes: "Ionian", "Dorian", "Phrygian", "Lydian", "Mixolydian", "Aeolian", or "Locrian"
- "Chromatic", "Whole Tone", or "Diminished"

```
buildScale("I")
buildScale("ii", octave=[2, 3])
buildScale("V7", octave=3, inversion=2)
buildScale("vi", octave=3, inversion=2)
```

```
transpose(pitchSet, origKey, newKey)
```

The **transpose** function accepts a pitch set in the form of a list of integers that are in one key and returns that set shifted to a new key. The **origKey** parameter is the original key that the pitch set belongs to and **newKey** is the key that will be transposed to; both **origKey** and **newKey** are specified using a string from the set of possible valid keys

```
transpose([60, 62, 67], "C", "G")
transpose([60], "Cm", "Fm") # returns [65]
```

Appendix C
Music reference

1 What is sound?

1.1 Waves

Sound is made up of waves of energy that travel through air or other physical media. Once those waves reach the human ear, they are captured by the outer ear and cause the eardrum to vibrate. These vibrations are funneled to a seashell-shaped muscle in the inner ear called the cochlea.

Frequency refers to the number of times a complete waveform passes through a single point over a period of time; this is how fast the wave is vibrating and is measured by cycles per second in a unit called hertz (Hz).

Wavelength refers to the physical length of one complete cycle of a wave in physical space, or the distance from one peak or zero crossing to the next. Wavelength has an inverse relationship with frequency: waves with a greater frequency have a smaller wavelength.

Amplitude is the measure of the height of a sound wave—or how high the peak of the wave is. This measurement is related to how we perceive the sound's loudness.

1.2 Frequency spectrum

Most of the sound we hear is made up of a combination of frequencies, called a frequency spectrum. We generally perceive the lowest frequency

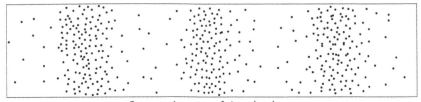

Compression wave of air molecules

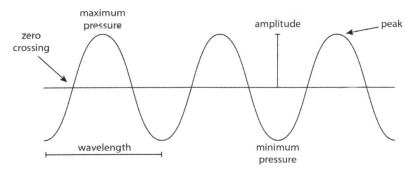

Figure A3.1 Parts of an audio wave.

as the pitch of a musical note. This frequency is known as the **fundamental** frequency, or often simply the fundamental. The remaining frequencies can be referred to as **partials** or **overtones**. This can be analyzed and visualized through a process called spectral analysis.

Partial frequencies that have an integer relationship with the fundamental frequency are considered harmonic. Otherwise, the partial is inharmonic.

1.3 ADSR envelope

A sound's envelope describes how the sound evolves over time. We can describe this in four simplified phases: **Attack, Delay, Sustain,** and **Release**.

The attack is the time from the onset to when the note reaches its maximum loudness. The decay is the time it takes the note to reach a secondary lower volume. The sustain is the actual loudness of this secondary volume. Finally, the release is how long it takes for the note to completely fade out.

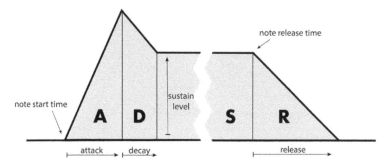

Figure A3.2 ADSR envelope.

1.4 Parameters of sound

Beyond the physical characteristics of a sound wave, we can break music into component elements: pitch, duration, timbre, loudness, texture, and spatial location.

Pitch refers to how high or low we perceive a sound. This is dictated by the sound's frequency. The higher the frequency, the higher we perceive a sound. Humans with full hearing can hear from 20 Hz to 20 kHz (20,000 Hz).

Duration refers to how long the sound lasts; this measurement relates both to rhythm and to a sound's envelope.

Timbre, or tone color, allows us to distinguish one sound from another. It is like the unique fingerprint of a sound. It is primarily determined by a sound's frequency spectrum and its envelope.

Loudness refers to how much energy a sound has. Loudness is very subjective, but can be expressed in decibels. Loudness is dictated by a sound's frequency, amplitude, and timbre. In digital music, this is related to gain. Gain isn't volume, but works as kind of a multiplier to audio's amplitude which can be adjusted to make the track more or less prominent.

Texture describes how different musical elements and instruments combine. A piece of music may have a dense texture if there are many instruments playing separate musical lines.

Spatial location refers to how we perceive a sound's physical origin location. In digital music, this can be controlled by an instrument's panning in a stereo mix, or what proportion of the track comes out of the left and right speakers. Producers can create more depth to a song and even replicate live recording by controlling the panning of individual tracks.

2 Rhythms

2.1 Beat values and notation

We can express a note's duration in units called **beats**. These durations are used in different combinations to create rhythmic patterns. Western music represents these durations using specific symbols:

Symbol	Name	Beats
O	Whole Note Larger open circle with no tail and no flag.	4
♩	Half Note Open circle with a tail and no flag.	2
♩	Quarter Note Solid circle with a tail and no flag.	1
♪	Eighth Note Solid circle with a tail and one flag or bar.	0.5 or ½
♪	Sixteenth Note Solid circle with a tail and two flags or bars.	0.25 or ¼
♩.	Dotted Half Note Open circle with a tail. The dot adds an extra beat.	3
♩.	Dotted Quarter Note Solid circle with a tail. The dot adds an extra half beat.	1.5
♪.	Dotted Eighth Note Solid circle with tail and one flag. The dot adds an extra quarter beat.	0.75

There are also symbols representing different durations of silence or "rests".

Symbol	Name	Beats
▬	**Whole Rest**	4
▬	**Half Rest**	2
⌇	**Quarter Rest**	1
⸜	**Eighth Rest**	0.5 or ½
⸝	**Sixteenth Rest**	0.25 or ¼

Notes are grouped together into repeating segments called **measures** or **bars**.

2.2 Time signatures

Time signatures describe how the rhythm of our song is structured. Time signatures are fractions. The top part of the fraction tells how many beats are in one bar. The bottom part of the fraction tells which note value is one beat.

The most common time signature is 4/4. It's so common that it's referred to as Common Time. Most of the music you have heard is likely in this meter. This time signature has four beats to each measure, with the quarter note receiving one beat. Other common time signatures are 2/4 and 3/4 time, which have two and three beats, respectively, with the quarter note receiving one beat. Cut time, or 2/2 time, has two beats per measure with a half note receiving one beat.

4/4	**Four-Four Time or "Common Time"** There are four beats in each measure, and each beat is a quarter note. This time signature is sometimes indicated using a special symbol:	**C**	♩♩♩♩
2/2	**Two-Two Time or "Cut Time"** There are two beats in each measure, and the beat value is a half note. Cut time is sometimes indicated with a "C" with a line through it.	**₵**	♩♩
2/4	**Two-Four Time** There are two beats in each measure, and the quarter note gets the beat.		♩♩
3/4	**Three-Four Time** There are three beats in each measure, and the quarter note gets the beat.		♩♩♩
3/8	**Three-Eight Time** There are three beats in each measure, and the eighth note gets the beat.		♪♪♪

2.3 Tempo

Tempo describes the speed at which the rhythm moves. One way to measure tempo is in beats per minute (BPM), meaning the number of beats played in one minute's time.

Below is a table with tempo ranges for common genres:

Genre	Tempo range (BPM)
Rock	100–140
R&B	60–80
Pop	100–132
Reggae	60–92
Hip-hop	60–112
Dubstep	130–144
Techno	120–140
Salsa	140–250
Bachata	120–140

3 Pitch

3.1 Intervals

Intervals are the difference in pitch between two sounds. In Western music, the smallest commonly used interval is called the half step, or semitone. This is the interval formed between two adjacent notes on a traditional keyboard. Two half steps combine to form a whole step. Western music theory uses a system of pitches with 12 notes to an octave with the distance between each pitch being a half step. Your ear will perceive two notes at an interval of an octave as the same pitch because the pitches are at a frequency ratio of 2:1.

Intervals are always measured from the lowest note. Intervals have two components: the generic interval and the quality. The generic interval is the distance from one note of a scale to another. The quality can be one of five options: Perfect, Major, Minor, Augmented, or Diminished. Each quality has a distinct sound and can generate different emotional responses. Here are some common intervals in music along with their frequency ratios and half steps:

Ratio	Interval name	Half steps
2:1	Octave	12
15:8	Major Seventh	11
16:9	Minor Seventh	10
5:4	Major Sixth	9
8:5	Minor Sixth	8

Ratio	Interval name		Half steps
3:2	Perfect Fifth	😇	7
45:32	Tritone	😈	6
4:3	Perfect Fourth		5
5:4	Major Third		4
6:5	Minor Third		3
9:8	Major Second		2
16:15	Minor Second		1
1:1	Unison		0

3.2 Notes

In Western music theory, notes are referred to through a lettering system: A, B, C, D, E, F, and G. These base pitches can also have accidentals that raise or lower the base pitch by a half step. This results in multiple names for the same pitch (A♯ is the same as B♭). When we reach G on a keyboard, the next note is an A as the system wraps around.

Each note can also be given a number which represents the octave.

3.3 MIDI

Computers use a system of numbers to represent pitch. Each note name corresponds to exactly one number. These numbers range from 0 to 127. Starting at 0, increasing the MIDI (Musical Instrument Digital Interface) number by one increases the pitch by one half step.

-1	0	1	2	3	4	5	6	7	8

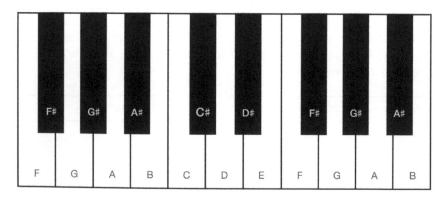

Figure A3.3 Natural notes on a keyboard.

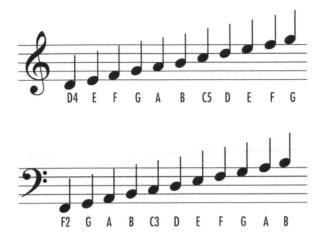

Figure A3.4 Natural notes for a Treble Clef (top) and natural notes for a Bass Clef (bottom).

C	0	12	24	36	48	60	72	84	96	108
C♯	1	13	25	37	49	61	73	85	97	109
D	2	14	26	38	50	62	74	86	98	110
E♭	3	15	27	39	51	63	75	87	99	111
E	4	16	28	40	52	64	76	88	100	112
F	5	17	29	41	53	65	77	89	101	113
F♯	6	18	30	42	54	66	78	90	102	114
G	7	19	31	43	55	67	79	91	103	115
A♭	8	20	32	44	56	68	80	92	104	116
A	9	21	33	45	57	69	81	93	105	117
B♭	10	22	34	46	58	70	82	94	106	118
B	11	23	35	47	59	71	83	95	107	119

3.4 Notation

To represent pitch graphically, we can use **musical staves** (or staffs). A staff is simply a set of lines. Each of the five lines corresponds to a specific musical note. The four spaces between a line also correspond to a note. The way we know what note corresponds to what line or space is through the use of a clef which indicates what notes are assigned to the staff. The two most commonly used clefs are bass clef and treble clef (see Figure A3.4). Bass clef is used most often for instruments with lower pitches; treble clef, for instruments with high pitches.

Each note can also have an accidental which raises or lowers the pitch of the note by a half step. A note with a sharp ♯ applied has its pitch raised by a half step, while a note with a flat ♭ applied is lowered by a half step. Accidentals carry for an entire measure, so subsequent repetitions of a pitch will also have the accidental applied. A natural ♮ cancels any other accidental for the remainder of the measure. When a note is beyond the range of what can be represented on a staff, extra lines—called **ledger lines**—can be drawn to continue the staff. The bass and treble staves can be combined to form something called the **Grand Staff**.

4 Scales

Scales are collections of notes which span one octave. They can be played one note at a time, ascending or descending from a starting note—or **tonic**.

The simplest scale is the chromatic scale, which includes all 12 notes in an octave. Each note is separated by an interval of a half step. Starting with a C on the piano keyboard, we would have the following notes:

$$C \; C\sharp \; D \; D\sharp \; E \; F \; F\sharp \; G \; G\sharp \; A \; A\sharp \; B$$

Or, using MIDI note numbers we could also write:

$$48, \; 49, \; 50, \; 51, \; 52, \; 53, \; 54, \; 55, \; 56, \; 57, \; 58, \; 59$$

4.1 Diatonic scales

Some of the most common scales in Western music are called diatonic scales. Diatonic scales consist of seven notes and include 5 whole step intervals and 2 half step intervals. The most common of these are the major and minor scales. There are 12 major scales and 12 minor scales, one for each possible starting pitch.

The major scale is made up of the following intervals: whole step, whole step, half step, whole step, whole step, whole step, half step. The major scale starting on C would have the following notes:

Note names	C	D	E	F	G	A	B	C
MIDI numbers	48	50	52	53	55	57	59	60
Intervals		WS	WS	HS	WS	WS	WS	HS

Minor scales also use seven notes out of each octave, but in a different order than major scales. Minor scales are often described as sad, melancholy, and distant. The minor scale starting on C would have the following notes:

Note names	C	D	E♭	F	G	A♭	B♭	C
MIDI numbers	48	50	51	53	55	56	58	60
Intervals		WS	HS	WS	WS	HS	WS	WS

4.2 Pentatonic scales

The pentatonic scales are a subset of the notes of the major and minor scales. There are just five notes in a pentatonic scale, and the scales have no half step intervals, which results in less dissonance between the notes. There are both major and minor pentatonic scales. The major pentatonic is created by omitting the fourth and seventh notes of the major scale:

Note names	C		D		E		G		A		C
MIDI numbers	48		50		52		55		57		60
Intervals		WS		WS		m3		WS		m3	

The minor pentatonic omits the second and sixth notes of the minor scale:

Note names	C		E♭		F		G		B♭		C
MIDI numbers	48		51		53		55		58		60
Intervals		m3		WS		WS		m3		WS	

4.3 Keys

Keys are the underlying organizational framework of most music and encode melodic and harmonic rules and conventions. The concept of keys is closely related to that of scales. Keys are composed of all of the notes in all of the octaves that make up the scale with the same name. For example, the notes in the key C major are the same as the notes in the C major scale across all octaves. But, while scales are usually played in increasing or decreasing order of pitch, the ordering of notes in a key doesn't matter. The notes that are part of a given key are called **diatonic**, and the remaining notes that are not part of that key are called **chromatic**.

Keys are organized according to the **Circle of Fifths** (Figure A3.5). The Circle of Fifths is essentially a pattern of intervals. Moving clockwise around the circle is moving the tonic note up by a fifth from the previous key. This adds one raised—or sharp—note as the seventh note of the scale. Alternatively, moving counterclockwise raises the tonic by a fourth and is often referred to as the **Circle of Fourths**. This adds one flat note as the fourth note of the scale.

5 Chords

Chords are a set of notes that sound at the same time. A chord's name comes from two parts: the root note and the quality. The first part is the root note of the chord, which is often the lowest pitch that sounds. The second part of

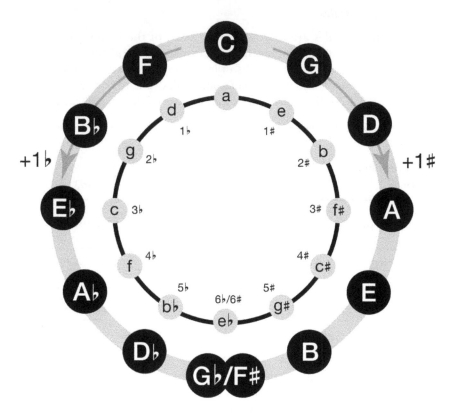

Figure A3.5 Circle of Fifths.

a chord's name is its quality. Each chord type has a consistent mathematical pattern. We can describe this pattern by naming the intervals for each note from the root note or by giving the number of half steps.

5.1 Triads

The most basic type of chord is the triad. Triads are made up of three notes: the root, the third, and the fifth. The interval between the first and second note is a third, as is the interval between the second and third; in other words, triads consist of two stacked thirds. Below are the different qualities of triads:

Quality	Intervals (name)	Intervals (half steps)
Major	root, Major 3, Perfect 5	0, 4, 7
Minor	root, minor 3, Perfect 5	0, 3, 7
Diminished	root, minor 3, diminished 5	0, 3, 6
Augmented	root, Major 3, Augmented 5	0, 4, 7

5.2 Seventh chords

Additional notes can be added to chords, which add more color. The most common type of extended harmony is the seventh chord. These chords have four notes instead of three. They are built on the triads we have already encountered, but with an additional note at the top.

This fourth note is a major or minor third away from the third note of the chord. Seventh chords can be built on major, minor, or diminished chords. Below are a few common seventh chords:

Quality	Intervals (name)	Intervals (half steps)
Major 7th	root, Major 3, Perfect 5, Major 7	0, 4, 7, 11
Dominant 7th	root, Major 3, Perfect 5, minor 7	0, 4, 7, 10
Minor 7th	root, minor 3, Perfect 5, minor 7	0, 3, 7, 10
Diminished 7th	root, minor 3, diminished 5, diminished 7	0, 3, 6, 9

6 Chord progressions

6.1 Roman numerals

A diatonic chord is any chord that can be played using only the seven notes of the current key. For example, if you're working in the key of C major, the diatonic chords consist of all of the chords you can play with only the white keys on the piano keyboard:

C D E F G A B

In C major, we would have the following diatonic chords:

CMaj Dmin Emin FMaj GMaj Amin Bdim

A system of Roman numerals can be used to refer to different diatonic chords by their scale degree rather than their specific name (i.e., Cmaj or Ddim). If the chord has a major quality, it gets an uppercase Roman numeral. If it has a minor quality, it gets a lowercase Roman numeral. Diminished chords are both lowercase and have an accompanied ° symbol. In music, only the numerals corresponding to numbers one through seven are used: I, II, III, IV, V, VI, and VII. That gives us the following Roman numerals for all of the diatonic chords of any key:

Scale degree	1st	2nd	3rd	4th	5th	6th	7th
Major keys	I	ii	iii	IV	V	vi	vii°
Minor keys	i	ii°	III	iv	v	VI	VII

6.2 Common progressions

Many early rock songs are built on blues progressions, such as I-IV-V. One of the most common progressions is the "doo wop" progression: I-vi-IV-V. The same chords can be reordered to form I-V-vi-IV. Hip-hop songs often use shorter progressions, frequently in a minor key with only one or two chords such as i-V, i-VI, and i-ii°. Below is a table of other common progressions:

Common major progressions	Common minor progressions
I – V – vi – IV	i – VI – VII
I – IV – V	i – III – iv – v
I – V – vi – iii – IV	i – VI – III – VII
I – vi – ii – V	i – v – VI
I – vi – IV – V	i – iv – v
I – iii – IV – V	VI – iv – v – i

7 Song structure

Song structure—also known as musical form—describes how musical ideas and material play out over a piece of music. One of the most basic units of song structure is the musical phrase. If melodies are paragraphs, then phrases are like musical sentences. They are complete thoughts that are punctuated and combined to form more complete and cohesive ideas. Phrases are often two, four, or eight bars in duration. These phrases are combined to form larger structures, which become the overall song form.

Repetition is an extremely important concept to both pop and classical music. One way in which a song might use this concept is through musical sections that repeat throughout a song. The most common way for musicians to classify and discuss the different parts of a song is to label them with letters. Every time we see a new section, we give it the next letter in the alphabet. If a section reappears, we reuse the letter we had already applied to that section. If a section reappears but is slightly different—maybe with different lyrics—we can give it a number. This works as a timeline of when a section first appears or reappears. If a song has three sections with no repeated sections, we can refer to that song's structure as ABC form.

In popular music, **verse-chorus** form is the most common song structure. At its simplest, this form is built on two repeating sections called the verse and the chorus. There are endless variations on this basic structure but this form tends to include:

Intro	A1 Verse	B Chorus	A2 Verse	B Chorus	C Bridge	B Chorus	Outro

8 Mixing and mastering

8.1 Effects

The use of mixing effects can give our track life and bring all of our elements together into a cohesive whole.

Dynamic range compression reduces the highest volumes and amplifies the lowest volumes, which shrinks the overall dynamic range of the audio. This ensures that the listener can hear the full range of volumes clearly.

Reverberation, or simply **reverb**, recreates the acoustical phenomenon created by sound waves moving through a physical space. Some of those waves bounce back to the listener. The waves that bounce back are heard as softer. Think of how sound reverberates through a concert hall, or even your bathroom. We can recreate this reverberation through applying reverb to our track.

The **chorus** effect is created by layering multiple copies of a sound, slightly offset from one another. It can have the effect of making audio sound richer and giving the illusion of having multiple performers.

The **phaser** effect is produced by playing multiple versions of the same sound together at the same time, but changing the frequency profile of each individual sound to get gaps or dips in the spectrum.

8.2 Filters

Filters allow us to shape the frequency spectrum of a selection of audio. Filters work kind of like gates that only allow certain frequencies to pass through. How filters behave is dictated by the filter's response curve. This is like a graph that describes which frequencies are allowed to pass through. There are four main types of filters: lowpass, highpass, bandpass, and notch.

Lowpass filters allow frequencies below a threshold—or cutoff frequency—to pass through while reducing the frequencies above the threshold. Highpass filters accept frequencies above the cutoff frequency. Bandpass filters reduce the frequencies above *and* below a band of frequencies; a bandpass is the equivalent of applying both a lowpass and a highpass filter. Notch filters are the opposite of a bandpass filter. Rather than bringing out a band of frequencies, a notch filter reduces the frequency band while all other frequencies pass through freely.

Index

Milton Keynes UK
Ingram Content Group UK Ltd.
UKHW031348071024
449327UK00033B/3048